BEGINNING SCIENCE
CHEMISTRY

R HART

Head of Science,
Churchdown Comprehensive School,
Gloucestershire

Oxford University Press

Oxford University Press, Walton Street, Oxford OX2 6DP

Oxford   New York   Toronto
Delhi   Bombay   Calcutta   Madras   Karachi
Petaling   Jaya   Singapore   Hong Kong   Tokyo
Nairobi   Dar es Salaam   Cape Town
Melbourne   Auckland

and associated companies in
Berlin   Ibadan

Oxford is a trade mark of Oxford University Press
© R. Hart 1985

First published 1985
Reprinted 1985, 1987, 1988, 1989
ISBN 0 19 914090 1

Cover photographs:
Front: ZEFA Picture Library
Back: Daily Telegraph Colour Library

Typeset by Rowland Phototypesetting Ltd
Bury St Edmunds, Suffolk
Printed in Great Britain by
Scotprint Ltd, Musselburgh

Contents

Introduction

Chemistry is all around you

The clothes that you are wearing, the food that you eat, and the air that you breathe are all part of chemistry. As scientists carry out research, they make new discoveries that affect your daily lives. You can read about just a few of them in the extracts that follow. These discoveries are being made now, while you are learning your chemistry. But remember that you may well be the scientists of tomorrow, making important discoveries of your own, so learn your chemistry well.

'Chemical engineers at Pennsylvania State University have been trying to solve a problem that has arisen with modern concrete bridges. It seems that concrete is about 15% air, and salty water from salt that has been put onto roads to clear ice, gets into the tiny holes, taking with it oxygen. This rusts the reinforcing iron that is inside the concrete. The engineers have been doing experiments to dry the holes out and then fill them with liquid plastic, which sets and keeps the salt out.'

'Carbon fibres – thin whiskers of pure graphite – are fifty times stronger than steel, but lighter than aluminium. Experiments have been going on for a number of years, mixing carbon fibres with more common materials like fabric and plastics to make new, super strong materials.'

'Protein extracted from certain jellyfishes have strange and very valuable properties. It seems that they can be made to glow when calcium and strontium compounds get near them. Doctors are researching these materials to see if they are of use as an early warning of heart disease. Sharks' livers also contain chemicals that can be used in anti-cancer drugs.'

'Polyethene terephthalate (polyester to you and me) has brought the fizzy drink manufacturers out of the Glass Age into the Plastic Age. Early plastics like polythene and PVC couldn't keep the carbon dioxide in drinks. It leaked out and glass bottle had to be used. Now, Terylene (one of the brand names of polyester) does the job.'

'Scientists in West Wales are going to pour two tonnes of lime into a lake to try to neutralise the ever increasing effects of acid rain.'

'Scientists reckoned that bacteria, cardboard, and even water used in manufacturing processes might be partly responsible for causing garments kept in polythene bags to go yellow. It is now thought that nitrogen oxide fumes in the air (mainly from car exhaust fumes) react with certain plastics to make a yellowing agent. The solution? Dip the finished fabrics in citric acid first of all before packing them and this seems to do the trick.'

'Dental scientists are developing a new white plastic that will soon replace metal amalgams used to fill holes in teeth. The plastic is put in in a soft form, but when specially coloured light is shone onto it, it hardens quickly and forms a hard substance that can do the chewing just as well as proper teeth.'

'"Dirty" industries like mining, smelting, and electroplating are pleased to hear that new bacteria are being bred that can withstand high concentrations of lethal cadmium and lead compounds in water, and which can be used to remove these impurities before they get into our drinking water.'

'Scientists have found ways of making superglue safe to use in surgery. They are using it to join delicate bone, blood vessels, and even nerves. They have had to remove dangerous cyanide groups from the superglue molecules but leave the bits that do the gluing.'

'Anorexia Nervosa – the slimming disease – could be caused by a zinc deficiency, scientists think. Zinc, like iron, is essential in our diet because it is necessary for the growth and repair of cells. Absence of zinc in our bodies can also depress our sense of taste and smell. Foods that contain good supplies of zinc are milk, meat, fish, and wholemeal bread.'

Methods

Useful information about chemicals, rules, and possible dangers is presented in these symbols.

1.1 Safety first

There are dangers all around you, but most of them you take for granted. You can't afford to do this in a laboratory.

Danger

First of all, let's look on the black side. If you were careless in the laboratory, you could poison yourself or even blow yourself up!

Now let's be realistic. These things could happen, but they could just as easily happen in other places. In fact laboratories are pretty safe when compared with other places in schools.

Figure 1 shows some statistics about accidents that happened in schools all over the country, during one school year. The Accident Scale in the second column compares how often accidents happened in different parts of the school. The higher the number, the more accidents there were. You can see that the most dangerous place appears to be the playground, and the safest place, the toilets!

Laboratory rules

But wait a minute. This doesn't mean that accidents will not happen in the laboratory. It means that science teachers and science pupils are generally careful people who try not to *let* accidents happen. They have rules about working and behaving in the laboratory.

Here are some rules for you to think about:

1 When you are in the laboratory:
      do not run,
      put your satchels, bags, and coats safely out of the way,
      if you have long hair, tie it up.

2 Always:
      wear safety specs, goggles, or a face mask when you are heating anything, or when you are doing any experiment that may be dangerous.

3 Never:
      eat or drink anything in the laboratory,
      look down into a test tube that contains chemicals,
      point a test tube at anyone while it is being heated,
      play with electrical switches,
      play with fire.

4 Always:
      ask your teacher if you are not sure how to do something,
      follow instructions carefully.

5 Never, ever:
      play around,
      make up your own experiments without first checking with your teacher.

6 Before you leave the laboratory:
      wipe the bench and tidy up,
      put your stool under the bench or out of the way.

place of accident	accident scale
playground	20.0
playing field	10.2
gymnasium	9.2
classroom	3.8
corridor	2.1
cloakroom	2.0
handicraft room	1.6
laboratory	1.3
swimming pool	1.1
domestic science room	1.0
stairs	0.8
toilets	0.4

Figure 1 This table compares how often accidents occurred in different parts of schools in one year. Accidents happened in the playground far more often than in the laboratory.

When working in a laboratory, you should always tie back your hair if it is long and wear goggles when you are heating anything.

Symbols

Many of the jars and bottles in the laboratory have safety symbols on them. Here are the more common ones, and some others you may also meet:

These symbols tell you about chemicals:

These symbols tell you what you must not do. They always have red circles and red diagonal lines:

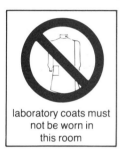   

laboratory coats must not be worn in this room	smoking, drinking, and eating prohibited	do not drink this water	fire, open light, and smoking prohibited

And these show what you must do. They have blue circles:

respiratory protection must be worn	head protection must be worn	eye protection must be worn

Exercises

1 Read this passage and see how many broken rules you can find:

Fred ran up the stairs and staggered through the door of the science laboratory. Falling over a stool that was in the middle of the floor, he threw his bag onto the bench and collapsed in a heap.

The teacher hadn't arrived for the lesson yet, so Fred got out his lunch and ate a sandwich. After this, he decided to start the lesson on his own. He had a recipe for gunpowder that he wanted to try out. So he looked around for the chemicals and mixed them in a heap on the bench. Nothing happened, so he decided they needed heating. He put them into a test tube, but couldn't find any matches to light the Bunsen burner. So he tore a page from his exercise book and lit it from the gas water heater. Being a tidy person, he then made a bonfire of all his rubbish in the sink.

The gunpowder didn't seem to be doing much. Fred gave it a poke with his pencil, and peered down the tube so as not to miss anything important. . . .

2 What do you think happened to Fred?

3

1.2 Apparatus

Chemists use special pieces of apparatus for special jobs.

Naming apparatus

You will use many different pieces of apparatus in your science lessons. You must learn to call them all by their correct names. You must also learn to spell their names correctly.

Most of the apparatus is made of borosilicate glass, often called *Pyrex*. Pyrex is expensive and although it is quite tough, it will break if dropped or mistreated. Take care.

Here are some of the pieces you will meet.

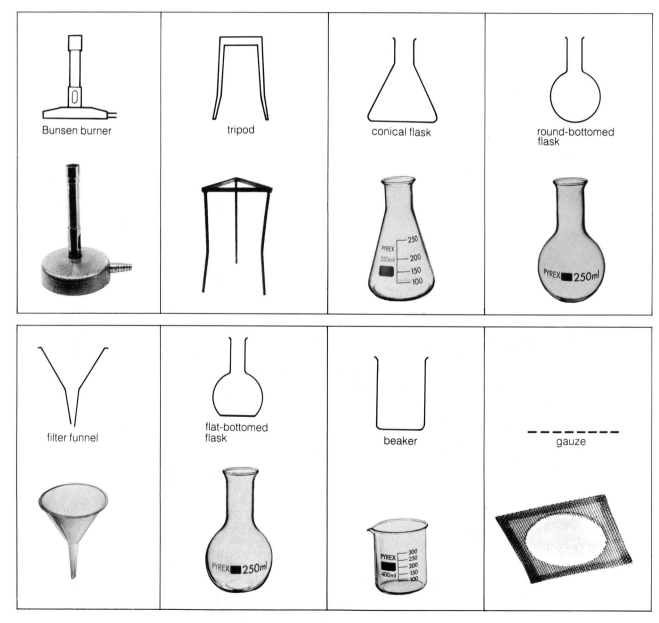

Bunsen burner

tripod

conical flask

round-bottomed flask

filter funnel

flat-bottomed flask

beaker

gauze

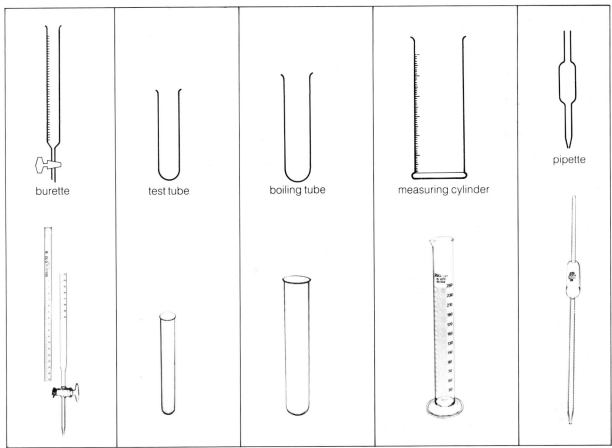

Figure 1   Some common pieces of laboratory apparatus

burette   test tube   boiling tube   measuring cylinder   pipette

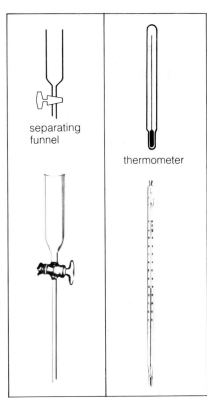

separating funnel

thermometer

Drawing apparatus

Chemists use diagrams instead of true drawings, to show apparatus. A diagram is shown above each photograph in Figure 1. Diagrams are simpler, and easier to draw. They look neater too. Remember these rules for drawing diagrams:

1   Use a pencil.
2   Use a ruler for all straight lines.
3   Label any piece of apparatus that is not easily recognised.
4   Label the substances in the apparatus.

Exercises

1   Learn to spell the names of the pieces of apparatus you have met. Your teacher may give you a test.
2   Practise drawing diagrams of these pieces of apparatus, for yourself.

1.3 Techniques (1)

Many pieces of apparatus are used for measuring.

Volume

The glassware you use will often have units of volume marked on it. These are the same units that are used to measure the volumes of everyday things like washing-up liquid, medicines, and soft drinks.

The units usually used are:

the *litre*, which has the symbol **l**
the *millilitre*, which has the symbol **ml**.

The container of washing-up liquid in Figure 1 has its volume marked in ml.

'Milli' means 'one-thousandth of', so you need 1000 millilitres to make one litre:

1000 ml = 1 l

Most countries sell petrol by the litre instead of by the gallon.

Two other units that are often used in chemistry are:

the *cubic decimetre*, which has the symbol **dm³**
the *cubic centimetre*, which has the symbol **cm³**.

Just like millilitres and litres:

1000 cm³ = 1 dm³ = 1 l

Don't be confused by these different units. Remember that:

a cubic centimetre is the same as a millilitre
a cubic decimetre is the same as a litre.

Beakers and conical flasks may have either cubic centimetres or millilitres printed on them. The beakers in Figure 2 both hold the same volume.

Figure 1 This container holds 750 ml of washing-up liquid.

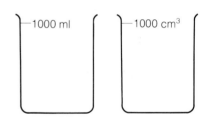

Figure 2 Each beaker will hold 1 litre of liquid.

Measuring volume

You can measure the volume of a liquid using a *measuring cylinder*. This is a glass container marked or *graduated* in either cm³ or ml. But there is a problem. When a liquid is poured into a narrow tube, its surface is not flat, but curved. This curve is called a *meniscus*. Look at Figure 3.

When you read the volume of a liquid in a measuring cylinder you must follow three rules:

1   Put the measuring cylinder on a flat surface.
2   Have your eyes at the same level as the surface of the liquid.
3   Take the reading from the bottom part of the meniscus.

The volume of the liquid in Figure 3 is 55 cm³.

The burette and pipette

You can measure out a liquid more exactly, using a burette or a

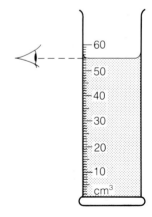

Figure 3 The surface of the liquid is curved. This curve is called a meniscus.

pipette. Figure 4A shows a *burette*. When the tap at the bottom is opened, the liquid will run out slowly. By noting the change in level of the liquid, you can tell how much has run out.

Figure 4B shows a *pipette*. Liquid is sucked into the pipette, using a safety filler, until the bottom of the meniscus just touches the line. You then know the exact volume of the liquid. For example, the pipette in Figure 4B holds exactly 25 cm³ when it is filled. The liquid is then run out into a beaker or flask.

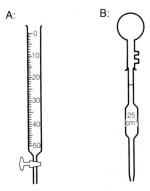

Figure 4   A: A burette. B: A pipette.

The thermometer

The thermometer is an instrument that measures temperature. Scientific thermometers are marked or graduated using the *Celsius scale.*

At normal atmospheric pressure, the freezing point of pure water is 0 °C (nought degrees Celsius) on this scale. Its boiling point is 100 °C. These are the two *fixed points* of the scale and there are 100 degrees in between. The thermometer in Figure 5 shows a temperature of 45 °C.

When you use a thermometer, remember to follow these rules:

1   Handle the thermometer carefully. It can easily roll off the bench and break.

2   Look at the scale first and make sure that you can read and understand it.

3   Keep the bulb of the thermometer in the substance while you are reading the temperature.

4   The mercury level will fall on its own, when the thermometer is removed from the substance. Do not cool it under the tap.

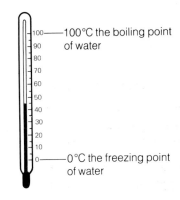

Figure 5   This thermometer shows a reading of 45°C.

Exercises

1   What is a meniscus? Draw a measuring cylinder with water in it, to help your explanation.

2   Write down the three rules for using a measuring cylinder. How much liquid is there in the measuring cylinder in Figure 6?

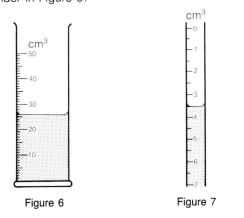

Figure 6

Figure 7

3   The burette in Figure 7 was full to begin with. How much liquid has been run out of it?

4   Has the pipette in Figure 8 been filled correctly? Explain your answer.

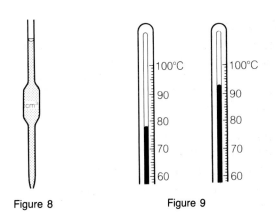

Figure 8

Figure 9

5   What temperature is shown on each thermometer in Figure 9?

6   How many cm³ are there in:
a) 1 dm³?   b) 1 l?   c) 0.5 dm³?   d) 0.25 l?

1.4 Techniques (2)

Chemists usually need to know how much of a chemical they have used.

Collecting gases

Quite often, you may need to collect a gas that is being made in a chemical reaction. Figure 1 shows apparatus for collecting the gas in a test tube.

The reaction takes place in a test tube or boiling tube. This is held in place by a *clamp* and a *retort stand*. The gas from the reaction travels along a *delivery tube*. It bubbles through the water and rises into a test tube. As it collects in the test tube, it pushes the water down until the test tube is full of gas. This is called *downward displacement of water*. If you think about it, you will see that this method would not work for a gas that dissolves in water.

Weighing

The amount of a substance is called its *mass*. Mass is measured in these units:

the *kilogram*, which has the symbol **kg**
the *gram*, which has the symbol **g**.

'Kilo' means '1000', so there are 1000 grams in a kilogram:

1000 g = 1 kg

At the supermarket, you can buy sugar in kilogram bags (see photo).

The mass of an object is usually measured on an instrument called a *top-loading balance*. Look at Figure 2. The number in the little window shows the mass of the beaker, to one decimal place. It is 48.5 g. Now look at the black bar beside the number. Its top edge meets the diagonal scale at 6, so 6 is the second place of decimals. The full reading is 48.56 g.

Top-loading balances are very sensitive and must be placed on a firm bench away from draughts.

Weighing out a powder

This sounds easy, but make sure you do it properly. First you must weigh the test tube or beaker that will hold the powder. Next, you must take the powder from the bottle carefully, without spilling any. Figure 3 shows the correct way to hold the bottle and spatula. Do not take too much powder at a time. Then, when the powder is in the test tube or beaker, you must weigh both together, and find the mass of the powder by subtraction. Set your results out neatly, as in this example.

Example

Mass of beaker	= 48.56 g
Mass of beaker + powder	= 72.06 g
Mass of powder	= 72.06 g − 48.56 g
	= 23.50 g

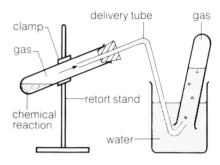

Figure 1 This is the apparatus used to collect a gas.

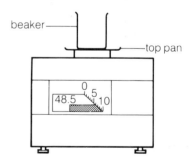

The mass of this sugar is 1 kg.

Figure 2 A top-loading balance measures the mass of an object.

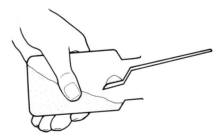

Figure 3 This is the correct way to take powder from a bottle.

8

The Bunsen burner

In 1852, Robert Wilhelm Bunsen became Professor of Chemistry at the University of Heidelberg, in Germany. One of the many important things he did there was to invent an apparatus that would mix gas and air in just the right amounts to give a hot, clean flame. The result was the *Bunsen burner*. See Figure 4.

When you use a Bunsen burner, remember these points:

1 Before you light it, close the air hole.

2 Get a lighted match or splint ready. Next turn on the gas and light it. Then open the air hole.

3 When the air hole is fully open, the flame is very hot and noisy, and nearly invisible. It is called a roaring flame.

4 The blue cone in the middle of the roaring flame is gas that has not yet burned. The hottest part of the flame is just above the blue cone.

5 When the air hole is closed, the flame is luminous and sooty. So always have the air hole open or half open when you are heating something.

6 If you are not using the Bunsen burner for a time, close the air hole so that the flame is luminous and can be easily seen. This may prevent an accident.

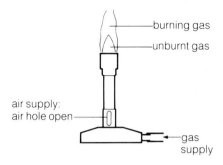

Figure 4 A Bunsen burner

Heating things in a test tube

When you are heating a solid in a test tube, remember:

1 Always use a test tube holder.

2 Always wear safety specs.

3 Always hold the test tube at an angle, as shown in Figure 5.

4 Never look down the test tube or point it at anyone.

When you heat a liquid in a boiling tube:

1 Don't put too much liquid in, or it will boil over.

2 Add a few anti-bumping chips to make it boil smoothly.

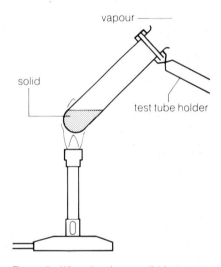

Figure 5 When heating a solid in a test tube, always hold the tube at an angle.

Exercises

1 a) Explain what is meant by the 'downward displacement of water'.
   b) Why is downward displacement of water unsuitable for collecting a gas that dissolves in water?

2 Write down the mass of each object shown below:

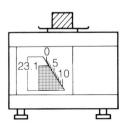
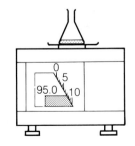

3 These results were obtained when a powder was weighed out into a test tube:
Mass of test tube   = 25.43 g
Mass of test tube + powder = 48.30 g
What mass of powder was put into the test tube?

4 Here is an experiment to try in the laboratory. Light a Bunsen burner and open the air hole to obtain a roaring flame. Hold a piece of wire in a pair of tongs, and put it into the flame in each of the positions shown. What can you tell about the temperature of the flame in the three positions?

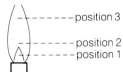

1.5 Separating mixtures (1)

When a chemist does an experiment, he often ends up with a mixture of substances instead of just one. He must know how to separate the substances.

Pure substances and mixtures

Every substance that exists is either a solid, a liquid, or a gas. For example, iron, cement, wood, and plastic are solids. Water, sulphuric acid, and petrol are liquids. The air, carbon dioxide, and hydrogen are gases. Solid, liquid, and gas are called the three *states of matter*.

Can you identify the three states of matter in this photo of a volcanic eruption?

A single substance that has nothing else mixed with it is called a *pure substance*. If there is anything else mixed with it, then it is a *mixture*. Figure 1 shows some everyday mixtures and the table below tells you what some mixtures are made of.

Figure 1 Coins, rivers, the sky, and soft drinks are all mixtures of substances.

mixture	states of matter involved	contents
the air	gases	mainly oxygen and nitrogen
the sea	solids, liquids, and gases	The sea is not pure water. It has salt, oxygen and all sorts of other substances dissolved in it.
fizzy drink	solid, liquid, and gas	sugar, flavouring, colouring, and carbon dioxide all dissolved in water.
brass	solids	copper and zinc
salad dressing	liquids	oil and vinegar. These liquids do not dissolve in each other and have to be shaken up

Chemists use different methods to separate different types of mixtures. Read on, and you will learn about them.

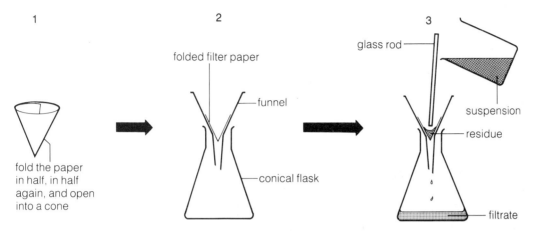

Figure 2   Filtering a suspension

Filtering

This method is used for separating a solid from a liquid in which it hasn't dissolved. For example, tea contains tea leaves. To separate them, you pour the tea through a strainer. The liquid passes through the holes in the strainer, but the leaves don't.

Dirty water contains tiny particles of dirt that have not dissolved. They just float about in the water. This type of mixture is called a *suspension*. The dirt is removed by pouring the water through a laboratory strainer called *filter paper*. Like a tea strainer, filter paper has holes in it, but they are too small to see without a microscope.

Figure 2 shows how to filter a suspension:

1   Fold the paper in half, and in half again. Open it out into a cone.
2   Put the cone into a filter funnel, and set this in a conical flask.
3   Using a glass rod as a guide, pour the dirty water into the paper cone. Do not overfill it.

The dirt that remains in the paper is called the *residue*.
The clean water that filters through is called the *filtrate*.

The centrifuge

Filtering is a good method for separating the liquid and solid in a suspension. Sometimes, however, it is too slow, or the amount of solid is so small that it would get lost in the filter paper. In this case, a *centrifuge* is used instead. Look at Figure 3. The suspension is put into a small test tube which is then placed in one of the buckets. The opposite bucket is counter-balanced, using a test tube of water. When the centrifuge is switched on, it spins at high speed, the bottoms of the buckets swing out, and the solid in the suspension is flung to the bottom of the test tube. The liquid can then be drawn off with a dropper, as shown in Figure 4.

Figure 3   A centrifuge is used to separate the solid from the liquid in a suspension. When the centrifuge is switched on, it spins at high speed; the buckets containing the test tubes swing out, and the solid is flung to the bottom of the test tube.

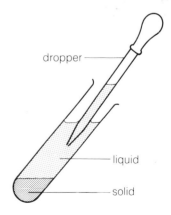

Figure 4   Draw the filtrate off with a dropper to separate the solid residue.

11

The separating funnel

In the table on page 10, you saw that salad dressing was made of oil and vinegar. These two liquids do not dissolve in each other. They are called *immiscible* liquids. But if they are shaken together they break up into tiny droplets, forming a mixture called an *emulsion*. If an emulsion is left to stand it will eventually separate again, into the two liquids. In this case, the oil will end up floating on the vinegar. This is shown in Figure 5. The two liquids can then be separated using a *separating funnel*. Look at Figure 6. When the tap is opened, the vinegar starts to run out. The tap is closed when all the vinegar has gone, leaving the oil behind.

Patience

One type of mixture that will separate itself, if you are prepared to wait long enough, is the *colloid*. A colloid consists of tiny particles of solid suspended in a liquid. But the particles are so small that they pass through a filter paper, and they are not heavy enough to be centrifuged. Often they are so small that they cannot even be seen with a microscope.

A good example of a colloid is starch in water. If a beam of light is shone through the liquid, as in Figure 7, it shows up like a beam of sunlight in a dusty room. This happens because the tiny starch particles disperse the light. If the liquid is left standing for several hours, the starch gradually settles out.

Figure 5 When an emulsion is left to stand, it separates into layers. In this case, an oil layer floats on a vinegar layer.

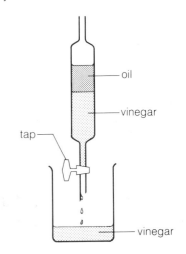

Figure 6 A separating funnel is used to separate immiscible liquids.

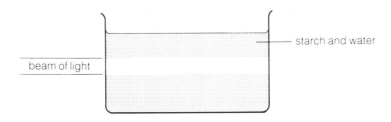

Figure 7 'Seeing' the colloid particles in starch and water.

Exercises

1 What are the three states of matter? Give one example of each.

2 Copy the table below, and write an example of a mixture in each empty box. For example, coffee is a mixture of solid and liquid. Try to find examples that have not been used earlier.

	solid	liquid	gas
solid		coffee	
liquid			
gas			

3 In your own words, say what each type of mixture is made of:
  a) an emulsion
  b) a suspension
  c) a colloid

4 Explain, with the help of a diagram, how you would use a separating funnel to separate cooking oil from water.

5 What do the words 'filtrate' and 'residue' mean?

6 What are immiscible liquids? Give two examples.

7 Suppose you own a scrap metal yard. Someone has brought in a load of iron mixed with aluminium. Can you think of a quick way to separate the iron from the aluminium?

1.6 Separating mixtures (2)

The mixtures on these pages are a little more complicated. The methods for separating them are very important and are used a great deal in chemistry.

Evaporation

When a solid mixes completely with a liquid, so that the solid disappears and a clear liquid is left, we say that the solid has *dissolved*. The solid is called the **solute** and the liquid is called the **solvent**. The clear liquid is called a **solution**. Sea water and sugar water are good examples of solutions.

To get a solid back out of a solution, a chemist boils the solution. The liquid boils away, but the solid is left behind. The same thing will happen, more slowly, if the solution is left open to the air in a warm place. The liquid turns to a vapour – it **evaporates** – leaving the solid behind. Figure 1 shows what remains when tap water is left to evaporate on a glass dish. The tap water must have contained dissolved solids. In hot parts of the world, salt is extracted from sea water by evaporation. (See photo on page 41.)

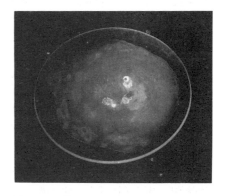

Figure 1 The solids dissolved in tap water are left behind when the water evaporates.

Distillation

Sometimes a chemist wants to get the liquid from the solution. For this, a method called **distillation** is used. Look at Figure 2. The liquid in the round-bottomed flask is a solution of salt in water. When it is boiled, the water turns to steam and leaves the salt behind. The steam travels along the delivery tube and into the test tube. By this time, it has cooled and turned back into water. This is called **condensation**. The water in the test tube will be pure.

The apparatus in Figure 2 wouldn't work for long, because the glass would soon get as hot as the steam, so the steam would no longer condense. This can be avoided by using a special condenser, as shown in Figure 3. It is called a **Liebig condenser**, after the man who invented it. The apparatus works in the same way as before. But this time the condenser is a tube with a glass jacket all the way around it. Cold water from the tap flows through the jacket, and keeps the tube cold. Figure 4 shows the condenser in more detail.

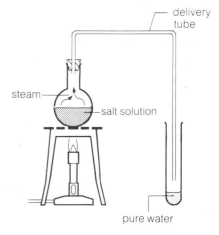

Figure 2 Simple distillation separates pure water from a solution.

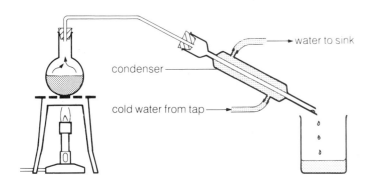

Figure 3 A laboratory distillation apparatus

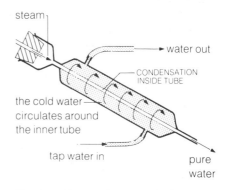

Figure 4 The condenser cools the steam so it condenses back to water.

Fractional distillation

Wine is a mixture of ethanol and water. A mixture of liquids like these can be separated by distillation if a *fractionating column* is used. Look at Figure 5. When the flask is heated, the ethanol and water both turn into vapours and rise up the fractionating column, which is packed with glass beads. The boiling point of steam is higher than the boiling point of ethanol, so the steam condenses on the glass beads more easily than the ethanol. The water runs back into the flask. The ethanol vapour goes on into the condenser where it turns back into liquid. You can read more about fractional distillation on pages 108 and 109.

Chromatography

If coloured ink is dripped onto blotting paper, it spreads out and sometimes separates into rings of different colours. This is because the ink is in fact a mixture of several different coloured inks. They separate because they soak into the paper at different speeds. This method of separation is called *chromatography*.

Suppose a chemist wants to investigate coloured substances like inks, or dyes extracted from plant petals. The method is this:

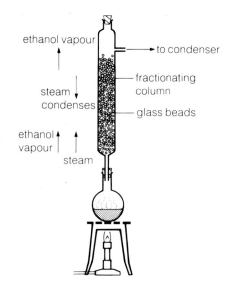

Figure 5 A fractionating column is used to separate two or more liquids.

A: the colours are spotted onto the paper

B: the paper is put into a jar of solvent

C: the paper is taken out of the solvent

Figure 6 This method of separating coloured substances is called chromatography.

1 The coloured substances are spotted onto a piece of chromatography paper and dried, as shown in Figure 6A.

2 The paper is made into a cylinder and fastened with clips. The cylinder is placed in a jar that contains a small amount of solvent (often water). A lid is put on the jar. Note that the spots of colour should be above the solvent level to start with, as shown in Figure 6B.

3 After some time, the paper is taken out and examined. The spots have separated into different colours. The solvent has carried the colours up the paper at different speeds, as shown in Figure 6C.

The paper with separated spots is called a *chromatogram*. A chemist can use a chromatogram to find out which coloured substances are in an ink or a plant extract. For example, the chromatogram in Figure 7 shows that the ink mixture was made of blue and red ink.

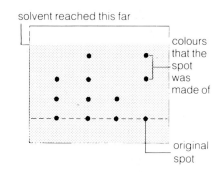

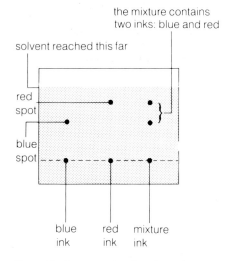

Figure 7 An example of a chromatogram

Chromatography is used a great deal in industry for analysing and detecting impurities. An example of one different sort of chromatography is described in the next section.

Gas–liquid chromatography

This is chromatography of a different sort but it still works in the same way. The substance under examination is injected into a long, thin tube through which a gas is flowing. The flowing gas is like the solvent soaking up the paper. In this case, the 'paper' is a chemical coated on the inside of the tube.

The different chemicals in the mixture under examination are carried along by the gas, but at different speeds, so they separate out. A detector at the end of the tube measures them as they come out and records them on a chart.

This sort of chromatography is used for isolating the chemicals in coal, for detecting poisons in a food sample, and even for finding how much alcohol there is in a driver's breath.

Figure 8 shows a chemist injecting a sample into a gas–liquid chromatography machine.

Figure 8 This chemist is injecting a sample into a gas – liquid chromatography machine.

Exercises

1 Imagine that you live in a very hot part of the world, by the sea. You have no salt deposits nearby, that you could mine. How would you obtain salt? Describe what you would do to provide enough salt for a large town.

2 This time, imagine that you are shipwrecked on an island in the Pacific Ocean. The island has no fresh water, and it hardly ever rains. There is, however, plenty of firewood and you managed to bring ashore with you some pots and pans and matches. Describe, with the help of diagrams, how you would obtain drinking water from the sea water.

3 A mixture contains two liquids, A and B. Liquid A has a boiling point of 78 °C and liquid B has a boiling point of 140 °C. The mixture is put into a flask. Draw the flask, and the rest of the apparatus that must be used with it to separate the two liquids. Label the apparatus.
Which of the liquids will boil off more easily when the mixture is heated?

4 A chemist did a chromatography experiment, to find out what inks were used in a mixture. Figure 9 shows the results she got. Which coloured inks did the mixture contain?

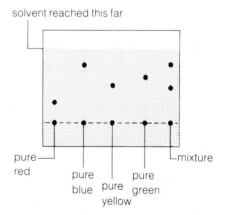

Figure 9

The story of a glass bottle

Glass – we take it for granted

So many things around you are made of glass that you probably take it for granted. Just think of windows, mirrors, milk bottles, ovenware, and all the laboratory glassware like test tubes, filter funnels, and beakers. They all had to be manufactured from glass.

The first glass objects were probably made by the Egyptians. They used glass jars and pots for holding perfumes, ointments, and oils. They made them by winding coils of molten glass, in the same way that children make clay pots from coils of clay. Later, men developed a method of blowing glass bottles from blobs of molten glass on the ends of metal pipes. The bottles were often shaped by the glass blower, using tools that could cut and flatten soft glass. Although craftsmen still make fine quality bottles and glasses by this method (look at Figure 1), most bottles and jars for everyday use are made by machine.

Figure 1   This man is a glass blower.

The ingredients for glass

Glass is made from sand (SiO_2), sodium carbonate (Na_2CO_3), and limestone ($CaCO_3$). When these three chemicals are heated together to 1500 °C they melt into a viscous liquid like molten toffee. When it cools, this liquid slowly hardens to form glass.

To help the ingredients melt, clean scrap glass is added. This scrap is called *cullet*. It is made of old bottles and jars that have been returned to the glass makers by breweries, shops, and dairies, because they are too chipped and scratched to use again. Just think how milk bottles get bashed about, for example. The old worn bottles become too difficult to clean, and too fragile as well.

At one time, most glass containers used for beer, soft drinks, and even jam, had a refund on them. So you could take them back to the shop when they were empty, and get a few pence in exchange. This ensured that the bottle owners got their bottles back to refill, and the glass factories got their cullet when the bottles became too old to use.

However, collecting old glass costs money. Manufacturers soon found that it was cheaper to make a *one-trip* bottle that was thrown away when empty. Some manufacturers even found that it was easier and cheaper to make containers out of aluminium, or plastic. The day of the ring-pull can and the 2-litre plastic bottle had arrived.

But meanwhile, many people were growing more concerned about litter and waste. They argued that it was a waste of fuel, to make containers that were thrown away after just one use. Besides, glass is difficult to dispose of, since it does not rot away, and it can be dangerous when it is left lying around.

Figure 2   Glass is collected from a bottle bank and recycled.

Because of pressure from these people, many manufacturers have started to take back empty bottles again. Some have put back deposits on their bottles. Some have left bottle banks for unwanted bottles and jars, in busy parts of towns. Look at Figure 2, for example.

Making the bottle

The glass ingredients are melted in a big furnace called a *tank*. The chemicals, which are now called the *batch*, are automatically fed in at one end of the tank. The batch is heated from the side by hot air, which is heated in turn by oil or gas burners. As the glass forms, it flows down the tank to the outlet, making room for more of the batch.

A big glass tank can hold as much as 2000 tonnes of glass. Once heated up, it is kept working continuously, night and day, for at least two years, until it needs to be broken down and rebuilt.

As the molten glass oozes from the end of the tank it is cut into small blobs called *gobs*, by metal shears. The gobs of glass fall into a heated metal mould (see Figure 3) and compressed air blows them into *blanks* or partly shaped bottles. The blanks are quickly transferred to another mould where more compressed air is blown in, forcing the bottles into their final shape.

Figure 3 Gobs falling into a mould. They fall so fast that they appear to the camera as a continuous stream.

All the time the glass is being moulded, it is kept hot by gas flames. The moulded glass must not be allowed to cool too quickly or it will develop stresses in its structure. These will make it weak and liable to crack easily. So the completed bottles are fed onto a conveyor belt called a *lehr*, where they are heated back up to a temperature of about 500 °C. They are then allowed to cool down very slowly as they move along. Look at Figure 4. Finally, when they are cold, they are checked for flaws and packed into boxes.

Figure 4 Completed bottles being fed onto a conveyor belt

Different kinds of glass

There are different kinds of glass for different purposes. Your daily milk bottle is made of *soda glass*. It is called this because of the sodium carbonate (soda-ash) which is used to make it. If the glass is clear, it is called *flint glass*. It could also be amber, green, or blue, depending on which additives are in it.

Iron compounds result in amber (brown) glass. Beer bottles are always made this colour. It disguises the fact that the beer itself may vary in colour from brew to brew.

Blue glass is made by adding cobalt oxide, and green glass by adding copper compounds. Some old red glass is very valuable because it contains gold compounds.

Some kitchen glassware is designed to withstand high temperatures, and sudden changes of temperature, without cracking. It is made of *borosilicate glass* or *Pyrex*. Pyrex contains boron oxide which makes its structure very strong. Most laboratory glassware, like flasks and beakers, is also made of Pyrex.

Very flat glass for shop windows is made by floating molten glass on huge beds of molten tin so that it is absolutely flat when it hardens. It is called *float glass*.

There are many other kinds of glass too, all containing different things. There is safety glass for car windscreens, and lead glass for expensive drinking glasses and for shields against radioactivity. There is optical glass for spectacle and camera lenses, and glass fibre for fire-fighters' suits and hulls of racing boats. So glass is not just a single simple substance. When you next say 'glass' stop and think a minute. Which type of glass do you mean?

Topic 1 Exercises

1 Say why it is dangerous to run, eat, or play in the laboratory.

2 Why is it important to put bags and satchels well out of the way in the laboratory?

3 Why must you tie up long hair in the laboratory, and wear spectacles or goggles when heating chemicals?

4 Why shouldn't you try out your own experiments without first asking your teacher?

5 Look at the safety symbols below. What do you think each one means?

6 Design your own safety symbol for a laboratory that uses a lot of hydrogen. Hydrogen is a gas which is very explosive when mixed with air. It can be set off by even tiny sparks. Does that give you any ideas?

7 Draw the apparatus you would use in the laboratory to collect a sample of a gas that has been made by heating chemicals in a round-bottomed flask. The gas has to be collected by the downward displacement of water.

8 Show how you would change the apparatus, if the gas in exercise 7 was soluble in water and lighter than air. (Hint: think about the downward displacement of air.)

9 Write down the names and symbols for the units that are used to measure:
    a) volume
    b) temperature
    c) mass.

10 Explain with a diagram the meaning of the word 'meniscus'.

11 Describe the steps you would take, the apparatus you would use, and the measurements you would make to find the mass of 15 cm³ of alcohol.

12 What is the difference between a burette and pipette? Use diagrams to help you explain.

13 Draw a thermometer with the Celsius scale. Find sensible values for the following temperatures and mark them on your diagram:
    a) your body temperature
    b) the temperature on a hot day
    c) bath-water temperature
    d) cold tap-water temperature
    e) the temperature of an ice/salt mixture.

14 What are the three states of matter?

15 Look at the diagrams below. What is wrong with the way each Bunsen burner is being used?

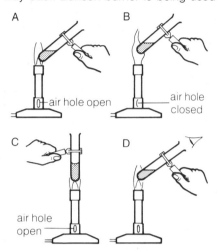

16 A solid was dissolved in water to make a solution. The solution was then boiled in a beaker until all the water had gone. The following measurements were made:

Mass of beaker   = 54.50 g
Mass of beaker + solution   = 99.63 g
Mass of beaker after boiling = 64.52 g

Find the mass of the solid and the mass of the water in the solution.

17 Which apparatus and which method of separation would you use, to get a pure sample of the first substance from each of the following mixtures?
    a) Copper sulphate and water. Copper sulphate dissolves in water.
    b) Tetrachloromethane and water. These liquids are immiscible and tetrachloromethane is heavier than water.
    c) Methanol and water. These liquids mix. The boiling point of methanol is lower than the boiling point of water.

18 Make your own colloid. Ask your teacher for dilute hydrochloric acid and sodium thiosulphate solution. Put the sodium thiosulphate solution into a beaker and add a few cm³ of the acid. After a few seconds, a hazy colloidal suspension of sulphur will appear and gradually darken. Try these tests on it:
    a) Try to filter it.
    b) Try to centrifuge it.
    c) Pour some of it into a large beaker and dilute it with water, until it is clear enough to see through. Now shine a strong beam of light from a projector through it, and see if the colloid particles show up.
    d) Leave the rest of it standing for a few days, and see if it separates itself.

Earth, water, and air

Earth, water, and air were all involved in the formation of these fossils in limestone.

2.1 The chemistry of the Earth

The study of the Earth's crust is called geology, but it is chemistry as well!

The Earth's structure

Look at Figure 1. If you could cut a segment out of the Earth, you would see that it is made of four quite different parts. The outer layer is called the *crust*. This varies in thickness from 10 km to 64 km at its deepest. It is made of huge *continental plates* that are slowly moving over each other, so that over millions of years the shapes of the land masses are very slowly changing.

The crust contains a wide variety of elements present as *minerals*. These are mainly oxides, sulphides, carbonates, and silicates of metals. Figure 2 shows the percentages of the different elements in the Earth's crust. You can see that by far the most abundant element is oxygen, and there is more aluminium than any other metal.

The next layer of the Earth is called the *mantle*. It is about 3000 km thick. It is very compact rock made of iron and magnesium silicates, and its temperature rises from 2000 °C at the top to 4500 °C at its deepest point. Near the surface, where the pressure is less, the rocks are molten.

The next 1600 km of the Earth is called the *liquid core*. This is a mixture of molten iron and nickel. The final and innermost part is called the *solid core*. This too is iron and nickel, but because of the very great pressure at the centre of the Earth, it is solid even though its temperature is very high.

The rocks of the Earth's crust

Igneous rocks The oldest rocks in the Earth's crust were once molten, and came from deep inside the Earth. The molten rock, called *magma*, spewed out in volcanic eruptions during the Earth's early life and solidified into hard rocks called *igneous rocks*. Examples are *granite* and *basalt*. You can see from Figure 3 that granite is a mixture of different chemical compounds. The black bits are mica (aluminium silicate) and the white bits are quartz (silicon dioxide) and Feldspar (potassium aluminium silicate). Granite is very hard and is used for buildings and tough things like kerb stones.

Sedimentary rocks Much later in the Earth's history, rocks that had been eroded by wind, rain, frost, and snow were swept down rivers and streams into the sea. There they settled in layers which slowly built up to form *sedimentary rocks*. Examples of these are *limestone* (calcium carbonate), *sandstone* (silicon oxide and calcium carbonate), *clay* (aluminium silicate), and *coal*. Limestone and sandstone are common building materials.

Sedimentary rocks often contain *fossils*, which are the imprints or petrified bodies of plants and animals that lived and died millions of years ago when the rocks were formed. Figure 4 shows a fossil in a piece of limestone.

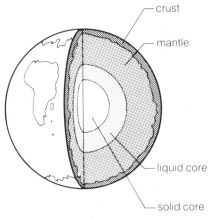

Figure 1 The Earth contains four different layers.

oxygen	47%
silicon	27%
aluminium	8%
iron	5%
calcium	4%
sodium	3%
potassium	2.5%
magnesium	1%
titanium	0.5%
others	2%

Figure 2 The percentages of the elements in the Earth's crust

Figure 3 Granite is a mixture of different chemical compounds.

Figure 4 This fossil is in a piece of limestone.

Metamorphic rocks Some sedimentary and igneous rocks were subjected to high pressures and temperatures at certain stages during the Earth's history. They turned into *metamorphic rocks*. Examples of these are *gneiss* (from granite), *marble* (from limestone), and *slate* (from clay and shale). Marble is white, often with streaks of colour due to impurities. It is a building material of great beauty. Slate is useful as a building material too, because it can be split into very thin sheets. Look at Figure 5.

Caves

Huge, underground caves are often found in parts of the world where the crust is made of limestone. Rain water is very slightly acidic because the carbon dioxide from air dissolves in it, making carbonic acid:

carbon dioxide + water → carbonic acid

When the acidic rain falls on limestone, it very slowly dissolves the rock away, forming a dilute solution of calcium hydrogencarbonate:

calcium carbonate + carbonic acid → calcium hydrogencarbonate

Caves form, where the rock has worn away. The dissolved calcium hydrogencarbonate is carried to rivers, and eventually reaches our taps. Tap water that contains calcium hydrogencarbonate is said to be *hard*. It does not lather easily with soap.

Figure 5 Slate can be split into very thin sheets and used for roof tiles.

Figure 7 These limestone fingers are called stalactites and stalagmites. Some of them are 250 000 years old.

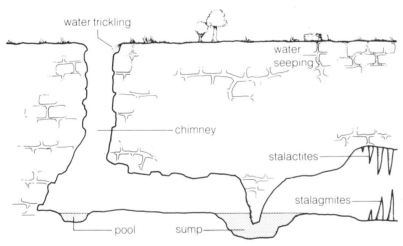

Figure 6 An underground cave system

Stalactites and stalagmites Figure 6 is a diagram of an underground cave system. At the end of the cave are formations called *stalactites* that hang from the roof, and *stalagmites* that grow up from the floor. These are caused by hard water. As it drips from the roof and falls on the floor, the water evaporates and deposits limestone. Over millions of years, these tiny deposits grow into long beautiful white fingers of limestone. An example from a cave in Wales is shown in Figure 7.

The same thing happens when hard water is boiled in a kettle. The *scale* or *fur* that forms in the kettle is made of limestone, just like stalactites and stalagmites. See Figure 8.

Figure 8 The scale in a kettle or pipe is calcium carbonate.

Minerals

The individual chemical compounds that make up rocks are called *minerals*. Minerals from which we extract metals are called *ores*. Examples of ores are haematite (iron oxide), pyrites (iron sulphide), and galena (lead sulphide).

Some minerals, like limestone and marble, are used as building materials, and gypsum (calcium sulphate) and clay are used to make cement. Many minerals are so beautiful that they are used for jewellery. But rubies are nothing more than aluminium oxide with impurities of chromium compounds, and sapphires are aluminium oxide with impurities of cobalt and titanium compounds.

Beautiful sculptures can be made from marble. This one by Rodin is called 'The Kiss'.

Figure 9 Diamond is nothing but carbon atoms arranged in a crystal lattice.

Probably the most valuable of all minerals is diamond. This is nothing more than carbon atoms arranged in a very strong crystal lattice. Look at Figure 9.

Exercises

1 What are the main parts inside the Earth? How deep is each layer and what does it contain?

2 Describe how igneous, sedimentary, and metamorphic rocks were made and give examples of each type.

3 What is hard water? How can you tell whether the water from your taps is hard? Do you live in a hard water area?

4 Coal and diamond are both minerals. They are both expensive. Can you explain why?

2.2 The air

The air around you is your life supply. It is also a valuable and important mixture of chemicals.

The atmosphere

The *atmosphere* is a mixture of gases that are held near the surface of the Earth by gravity. The mixture is thickest at the Earth's surface, but the higher you go, the thinner it gets. Look at Figure 1. The first 11 km of the atmosphere is called the *troposphere*. This layer contains most of the air, and the lower part of it has the clouds, formed by water vapour. Above 6 km the air is so thin that mountain climbers have to carry oxygen to breathe.

Voyager-1 leaving its launch pad on a mission to the outer planets. Because it travelled beyond the stratosphere it had to carry its own supply of liquid oxygen.

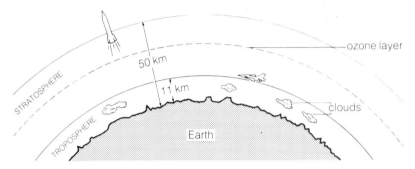

Figure 1 Our atmosphere consists of two main layers.

Above the troposphere is the *stratosphere*. It reaches up to about 50 km above the earth's surface. In the middle part of it is the *ozone layer*. This layer is very important. It prevents much of the dangerous ultra-violet radiation that comes from the sun from reaching the Earth's surface. Too much ultra-violet radiation can burn your skin very badly. Despite what many people think, there is very, very little ozone in the air at the earth's surface, even at the seaside. That's just as well because it is a very poisonous gas.

Jet aeroplanes can fly in the lower part of the stratosphere, but no higher, because there is not enough oxygen for them to burn their fuel. Rockets which go beyond the stratosphere must carry their own supply of liquid oxygen.

What's in the air?

Pure air contains two important gases. These are *oxygen* and *nitrogen*. In addition, there are smaller amounts of carbon dioxide, water vapour, and argon. There are minute amounts of neon, helium, krypton, and xenon, and even tinier amounts of hydrogen, ozone, and methane. Figure 2 shows the percentages of the main gases present in the air.

Other gases are present in the air as *impurities*. Some of these are due to car and lorry engines. When petrol is burned, carbon monoxide, nitrogen oxide, and lead dust are ejected into the air. Generally, these poisonous substances just spread out through the air, but in areas where there is a lot of traffic, they can build up to

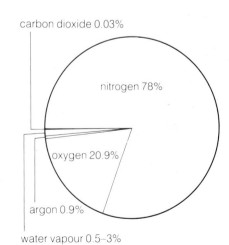

Figure 2 The percentages of the main gases in the air

dangerous levels. Lead is particularly harmful to young children. In some places, thick fogs called *photochemical smogs* can be formed when the impurities get trapped near the surface of the Earth. Look at Figure 3.

Some of the impurities in air come from factory chimneys. When oil or coal is burned in factories, soot and sulphur dioxide may be produced. The sulphur dioxide reacts with rain to form very dilute sulphuric acid. This falls as *acid rain*. It erodes buildings made of limestone, and it damages crops. It can also kill trees and fish.

When we burn fuel, we always produce *carbon dioxide*. Some scientists believe that we are burning too much fuel; if we continue, the increased carbon dioxide in the atmosphere will act as a blanket round the Earth, heating up its surface. A rise of even one or two °C would be enough to make the polar ice caps melt, and that would cause huge floods.

Figure 3 Car fumes and smoke from factories can cause air pollution.

What the air's gases are used for

The two main gases in the air have a number of very important uses. *Oxygen* is essential to animals for breathing. Deep-sea divers and high-altitude pilots need to carry their own supplies of oxygen to survive, and some patients in hospital need their own supplies too. Oxygen is also used to turn iron into steel. High temperature cutting and welding torches use oxygen. Look at Figure 4.

Nitrogen is used to make ammonia and this in turn is used to make nitric acid. From these two chemicals, plastics, drugs, dyes, fertilizers, and explosives are made. Liquid nitrogen at −200 °C is used to cool down air, in order to freeze peas and fish fingers ready for your freezer.

Although there is not much of them, argon and helium are important too. *Argon* is a very unreactive gas. It is used in light bulbs to stop the tungsten filament burning or evaporating. It is also used when metals like aluminium and titanium are being welded, to prevent them from oxidising. *Neon* is used in advertisement signs, as shown in Figure 5.

Figure 4 Oxygen is needed to produce this flame.

Figure 5 Neon is used in advertisement signs.

Combustion

When things burn, they combine with oxygen to form compounds called *oxides*. The Bunsen burner is a good example. In it, natural gas (methane) is mixed with air (which contains oxygen) and they react together when given energy in the form of a spark or flame. Methane is a *hydrocarbon*. This means its molecules are made up of hydrogen and carbon atoms. When it reacts with oxygen, the carbon turns into carbon dioxide and the hydrogen turns into hydrogen oxide (steam):

methane + oxygen → carbon dioxide + steam

The flame that is produced is nothing more than a very hot mixture of carbon dioxide and steam. If the air hole on the Bunsen burner is closed the flame becomes yellow and luminous, as shown in Figure 6. This is because some of the carbon is not completely burned and appears in the flame glowing brightly.

Figure 6 When the air hole is closed, the flame is yellow and luminous.

You need a flame or spark to light a Bunsen burner. But once it is lit, it will burn on its own because the reaction between methane and oxygen gives out heat energy.

Fire

Fires are burning substances. For a fire to continue, three things are needed. There must be something to burn (a *fuel*), *oxygen* (usually from air), and *heat energy*. When a fireman puts out a fire, he does so by removing one or more of these three things. He may cut off the fuel supply by turning off gas taps or cutting away things like wooden window frames. He may cut off the air supply by covering the flames with sand or foam or a blanket. He may cool the fire down by covering it with water. In each case, he is stopping a chemical reaction. Look at Figures 7 and 8.

Figure 7 The fireman is stopping this fire by covering it with water to remove the heat energy.

Figure 8 In each case, the fireman is cutting off the air supply. He is covering the flames with foam (left), a blanket (centre), and carbon dioxide (right).

Exercises

1 How high can you climb without taking your own supply of oxygen with you?

2 What does the ozone layer do?

3 Give two uses of oxygen and two of nitrogen.

4 a) Why is argon used in light bulbs?
   b) Which gas is used in advertisement signs?

5 What two gases are formed when natural gas burns?

6 a) What three things are needed for a fire to burn?
   b) Are these three things present in a Bunsen flame? Explain your answer.

7 What is a hydrocarbon?

2.3 Water for life

Each year, millions of tonnes of rain fall on the Earth. Most of the rain flows to the sea. Some of it we collect and make dirty before we send it on its way.

The water cycle

Most of the water we use is used over and over again. Every day, many tonnes of water are evaporated from the sea by the sun's heat, and turned into clouds. Look at Figure 1. The clouds drift over high land where they make rain. The rain falls onto the land and makes rivers and lakes. Some water soaks into the ground and fills wells and forms springs. Most of it runs back into the sea, where the whole process – called the *water cycle* – begins again. But we take some of the water from rivers, on its way to the sea, and use it for washing, drinking, cooking, making beer, making steel, and a thousand-and-one other things. Look at the information which follows.

It takes:

    5 litres of water to make half a litre of beer

    10 litres to make half a kilogram of chocolate

    65 litres for your bath

    70 litres to make a car tyre

and 200 000 litres to make a tonne of steel

Pollution

Rain water is generally pure – although not if it falls in a smoky, industrial area. But as soon as it hits the ground, it starts to dissolve things. It dissolves rocks, and soaks chemicals out of the soil, but these do not generally spoil it. In fact the dissolved substances often give water a pleasant taste. (Have you ever tried tasting pure, distilled water?) In some places the water is thought to be so good for you that it is bottled and sold! Look at Figure 2.

Our tap water was once rain water that drained into streams and rivers. When we use it, we add a great many more substances to it, and make it very dirty. It gets washed in, cooked in, and flushed down the lavatory. It eventually finds its way back to the river and sea, but only after it has been treated at sewage works. Even then it is not perfectly clean.

A certain amount of dirty water shouldn't bother a river too much, because the river has a way of cleaning itself up. It contains microscopic animals that can break down rotting vegetation and certain types of dirt. Green plants in the water provide the oxygen for this process, as well as for keeping fish alive. However the balance can be upset if the plants and fish are killed by poisonous chemicals, or if the plants grow too vigorously because of fertilizer that has drained into the river from farms. Water that can no longer keep itself clean is *polluted*. Look at Figure 3. This water is polluted by detergent.

Figure 1 Every day many tonnes of water evaporate from the sea and form clouds. This process is an important part of the water cycle.

Figure 2 Bottled water tastes good because of the substances dissolved in it.

Figure 3 This river is polluted by detergent.

Sewage

Nearly all the water in Britain that drains from roads and gutters, and all the water that people tip down the sink and bath hole, finds its way into the *sewers* and is led off to the *sewage works*.

At the sewage works, it is put through *screens* which are wire-mesh filters. These remove any rags and bits of wood and plastic. Next, it is led slowly through narrow channels so that grit and sand can settle out. Most of the impurities that remain are in the form of suspended solids. Then it is piped into a big *sedimentation tank* where it is slowly stirred. Here most of the suspended solids sink to the bottom as *sludge*. The sludge is led into big tanks called *digesters* where it heats up naturally and harmful bugs are killed off. It is then dumped on the land or at sea. The fairly clear water from the sedimentation tank is taken to a *biological filter*. Look at Figure 4.

In a biological filter, the water is sprayed onto a bed of stones where microscopic animals live. These feed on the *pathogens*, the dangerous bugs in the water.

Finally, the water is put through a *sand filter* before it flows back to the river. It is now clean enough for the river to deal with by its own natural processes.

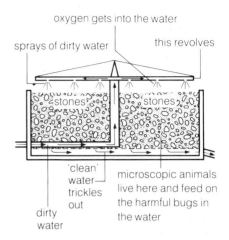

Figure 4 A biological filter in a sewage works

Drinking water

The water supply for your town or village probably comes from a reservoir or a river. A *reservoir* is a store of clean water, usually drained from hillside springs and streams. If river water is stored, it is taken from a clean part of the river, upstream from any sewage works or factories. Despite this, the water must still be treated at the *water works* before it can be consumed.

At the water works, the water is first put through screens, just like at the sewage works. Next it is sprayed through the air from jets, or let tumble down a tower. In this process oxygen is absorbed, and unpleasant smells and tastes removed. Next a chemical called *alum* may be added. This helps any fine particles of silt to cling together or *coagulate*, so that they settle out. Then the water is filtered through sand. It now looks and smells clean, but may still contain harmful pathogens. So a small amount of chlorine gas is dissolved in it, to kill them. Finally it is piped to a water tower or covered reservoir, where it waits for someone to turn on the tap. Look at Figure 5.

Figure 5 This water tower contains stored, clean water.

Exercises

1 Why is the water cycle called a cycle?

2 Describe the journey of a water molecule from the moment it leaves the sea to the moment it joins it again. Try to think of all the places it might visit.

3 List some of the types of pollution that may be found in a river. If there is a river near you, think of all the things that drain into it.

4 At the sewage works, what is the purpose of:
  a) the screens?
  b) the sedimentation tank?
  c) the digester?
  d) the biological filter?

5 Why is chlorine put into tap water?

6 What does 'coagulate' mean?

2.4 Water in the laboratory

Water dissolves things. That will be obvious to you if you take sugar in your tea, or swallow a mouthful of sea water.

Solutions

Solutions are mixtures of soluble solids, called *solutes*, and liquids, called *solvents*:

solute + solvent → solution

One of the best solvents is water. Solutions made with water are often called *aqueous solutions*. So a solution of copper sulphate in water can be described as $CuSO_4(aq)$. The 'aq' stands for aqueous.

Two questions that are often asked in the laboratory are:

1 Does this water have anything dissolved in it?

2 Does this solid dissolve in water?

Here is the way to find the answers:

1 If the water is heated until it evaporates, then any solute will be left behind as a solid residue. Figure 1 shows the apparatus that could be used for this. If a residue is left in the evaporating dish, then the liquid was really a solution.

2 Put some water in a test tube. Add a very small amount of the solid and stir carefully with a glass rod. Hold the test tube up to the light. If all the particles of solid have gone and a clear liquid is left, then the solid was soluble. Look at Figure 2.

Solubility

The amount of solute that will dissolve in water is called the *solubility* of that solute. Try to answer these two questions:

1 Is it easier to dissolve a solute, such as sugar, in a little or a lot of water?

2 Is it easier to dissolve it in cold or hot water?

The solubility of a solute depends on two things, the *amount of water* and its *temperature*. You probably gave these answers to the questions above:

1 It is easier to dissolve a solute in a lot of water.

2 It is easier to dissolve it in hot water.

Here is an example of the solubility of a substance:

The solubility of potassium nitrate in water is 104 g of potassium nitrate per 100 g of water at 60 °C.

A solution that contains 104 g of potassium nitrate in 100 g of water at 60 °C is called a *saturated* solution. It contains as much dissolved solid as it can hold at that temperature. How do you know this? Well, add some more solute. If the solution really is saturated, the extra solute will not dissolve. See Figure 3.

What happens when the temperature changes?

Look at Figure 3 again. In the second beaker, more solute was

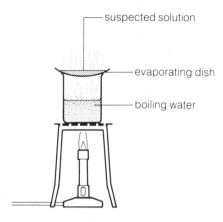

Figure 1 If anything is dissolved in the water, it will remain in the evaporating dish as the water evaporates.

suspected solution
evaporating dish
boiling water

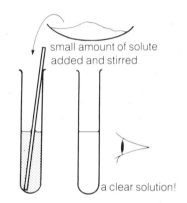

Figure 2 If a solid is soluble in water, it forms a clear solution when added to water.

small amount of solute added and stirred
a clear solution!

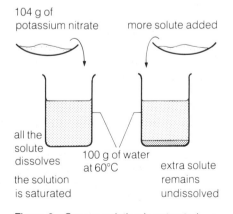

104 g of potassium nitrate
more solute added
all the solute dissolves
100 g of water at 60°C
the solution is saturated
extra solute remains undissolved

Figure 3 Once a solution is saturated, extra solute will not dissolve.

added to the saturated solution but it wouldn't dissolve. If the beaker is now heated, the extra solute will dissolve. Hot water can dissolve more solute than cold water can. *The solubility increases as the temperature increases*.

The opposite happens if the saturated solution in the first beaker is cooled. Cold water can dissolve less solute than hot water, so some of the dissolved solute will crystallise out. *The solubility decreases as the temperature decreases*. Look at Figure 4.

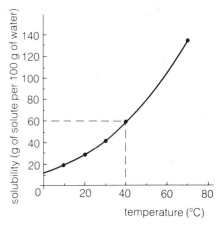

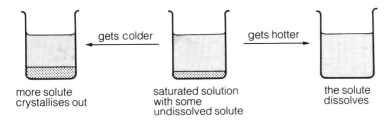

Figure 4 Solubility changes as the temperature changes.

Solubility curves

You can see that a solute has different solubilities at different temperatures. This information can be shown on a graph called a *solubility curve*. Look at Figure 5. The graph shows how the solubility of potassium nitrate increases as the temperature is raised.

Gases too?

Gases as well as solids dissolve in water. For example, fresh water contains a small amount of oxygen. When a beaker of water is heated, small bubbles appear on its sides long before the water reaches its boiling point. Look at Figure 6. The bubbles are dissolved oxygen coming out of solution as the water heats up. Unlike solids, gases are *less* soluble in hot water than in cold. Think of tropical fish. They need to have air bubblers in their tanks to keep their warm water supplied with oxygen so that they can breathe.

Gases can be made to dissolve better if they are pressurised. Carbon dioxide dissolves in water if it is put in under pressure. Think of fizzy drinks. As soon as the pressure is released, the gas comes bubbling out of solution.

Figure 5 A solubility curve for potassium nitrate. Using the graph, you can do two things:
1. You can find the solubility of the solute at a particular temperature. So at 40°C, the solubility of potassium nitrate is *60 g per 100 g of water*.

2. You can find the temperature at which a certain mass of solute will dissolve. So 60 g of potassium nitrate will dissolve in 100 g of water *at 40°C*.

Figure 6 When fresh water is heated, the oxygen dissolved in it bubbles out of solution.

Exercises

1 Explain what is meant by the terms 'solute' and 'solvent'.

2 You have a beaker of pure water and a beaker of clear river water. How can you tell which is which?

3 How would you find out if powdered chalk was soluble in water?

4 On what two things does the solubility of a solute in water depend?

5 Does the solubility of solids in water increase or decrease with increasing temperature?

6 Look at the solubility curve in Figure 5. What is the solubility of potassium nitrate at 50 °C?

7 Why are tropical fish tanks more likely to need air bubblers in them than cold-water tanks?

8 What would you expect to see if you warmed up a fizzy drink?

Chemistry and your teeth

Why brush them?

The first and simplest answer to this question is that your mouth would be pretty horrible if you didn't. Bits of food would collect in the gaps between your teeth. They would go bad and start to smell. The teeth themselves would become stained and discoloured.

Speaking biologically, your teeth become coated naturally with a substance call *plaque*, which contains bacteria. You can't see the plaque, but it shows up if you eat *disclosing tablets*, which stain plaque red. Look at Figure 1.

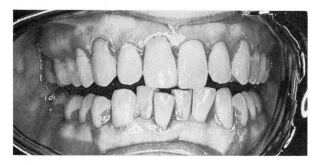

Figure 1 The stains show where the teeth are coated with plaque.

The bacteria in plaque are themselves harmless. But as soon as they come into contact with sugars, they quickly convert the sugars into acids such as *lactic acid*. These acids attack the *enamel* of your teeth and eat holes in it. Look at Figure 2. Once the enamel has been penetrated, bacteria get into the soft *dentine* under the surface, and decay starts. Dentists call dental decay *caries*. In addition, plaque can push back the gums from the top of the teeth, exposing the dentine there. The gums can become very inflamed and your teeth may even drop out.

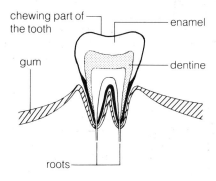

Figure 2 The main parts of a tooth

It doesn't take long for the acids to be formed. Within minutes of your first mouthful of food, the bacteria are in action, working on the sugar in any bits that have stuck to your teeth. Our modern diet is generally good for these bacteria and bad for teeth, since it includes a great deal of sugar. In fact, we each eat on average about a kilogram of sugar a week. Much of this is hidden in convenience foods like tinned soups, tinned beans, sauces, and even sausages and frozen hamburgers.

After reading that, you may not be surprised to learn that 99% of all the people in Britain have had dental treatment at one time or another. One third of the people in this country over the age of 16 have no teeth of their own!

The dentist

If you do have dental decay, the dentist has to use chemistry to mend your teeth. After drilling out the decayed dentine, he fills the cavity with a soft *amalgam*. This is a mixture of silver and tin dissolved in mercury. When silver and tin are mixed in the proportions of about 70% silver and 26% tin, and dissolved in mercury, they form a mixture which sets to a hard solid in about an hour. While it solidifies it expands a little, so that it properly fills the cavity the dentist has drilled.

A small amount of copper is added to dental amalgam, to make it even harder. A little zinc is added also, as a *scavenger*. It stops the other metals from reacting with the air. Sometimes, very small amounts of gold or platinum are added too. And recently, scientists have been experimenting by adding calcium fluoride. The reason for this will become obvious, when you read on.

Toothpaste

To stop your teeth from decaying, you must remove the plaque as often as possible. For most people this means cleaning teeth night and morning, and after lunch if possible.

If you lived in certain other parts of the world you would clean your teeth with a twig or a piece of wood, and salt or ashes. If you were a Roman, you would have used a mixture of ashes of oxen hooves, burnt egg shells, and powdered pumice, all flavoured with myrrh. As it is, like most of the people in this country, you probably buy a tube of toothpaste from the local supermarket. You probably get through about half a litre of the stuff a year.

Modern toothpaste doesn't really clean teeth any better than salt and ashes do, but it certainly speeds the process up. It contains a *polishing agent* like chalk and a small amount of *detergent* such as sodium lauryl sulphate. This is often blended with glycerol, and flavouring and sweetening (really!) are added too. In addition a *preservative* is added to stop the mixture going off, and a *humectant* to stop it from drying out. The whole lot is blended in a vacuum to make a smooth mixture. We probably all use too much of it on our brushes. One expert claims that most of it comes out again with the first spit! Look at Figure 3.

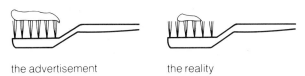

the advertisement   the reality

Figure 3 How much toothpaste should you use?

The big difference between ashes and toothpaste is *fluoride*. The British Dental Association claims that fluoride compounds can reduce the amount of dental decay by 15 to 20%. Fluoride strengthens the enamel and makes it more resistant to attack from acids. It has most effect on children, whose teeth are still growing. In other words, you.

Only a small amount of fluoride is needed. It can be taken in tablet form, or painted directly onto the teeth, or added to the public water supply. The last method is a controversial one: not everyone wants to drink fluoride. One simple method is to put it in toothpaste that people can buy if they want.

The fluoride compounds used in toothpaste are mainly tin(II) fluoride, sodium monofluorophosphate, and sometimes sodium fluoride. Each has exactly the same action, although the different manufacturers will always claim that their toothpaste is best.

Toothbrushes

The range of toothbrushes in the chemist's shop can be very confusing. They appear to come in every size, shape, hardness, and colour. However, dentists seem to agree that the best toothbrushes have nylon bristles (called *filaments*) with rounded ends. The filaments are very thin so that they can reach into every crevice and gap as well as brush the surface of the teeth and gums. In fact, a good toothbrush has more than 1000 filaments grouped together in small tufts.

The handle of the brush is important too. It must be made in such a way that it is easy to hold when wet, and comfortable to hold at the correct angle for cleaning. Handles are made of plastics such as cellulose ethanoate or styrene acrylonitrile. These are both *thermoplastics*. When they are heated, thermoplastics become soft and mouldable, just like toothpaste! The hot plastic is injected into a mould the shape of the brush handle, and then cooled so that it solidifies. Small holes are left in the head, ready to receive the filaments. Look at Figure 4.

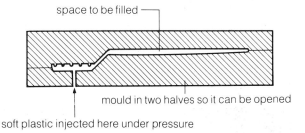

space to be filled

mould in two halves so it can be opened

soft plastic injected here under pressure

Figure 4 A toothbrush is made by injection moulding.

The filaments are made from another thermoplastic called nylon 6.12. They are made as long threads, by a process called *extrusion*. In this process, the nylon is heated until soft, and then squeezed through small holes in a device called a *spinneret*. Look at Figure 5.

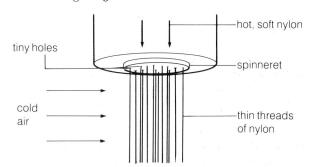

hot, soft nylon

tiny holes

spinneret

cold air

thin threads of nylon

Figure 5 Nylon filaments are made by extrusion.

The nylon threads are then chopped into lengths twice as long as the bristles will be. A machine gathers them together in bundles of 50 or more and folds them double. It pushes them into the holes in the toothbrush head, and fixes them in place with a small metal pin. Finally, the ends of the filaments are rounded off. Look at Figure 6.

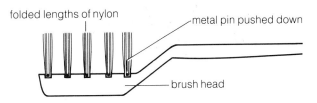

folded lengths of nylon

metal pin pushed down

brush head

Figure 6 A metal pin holds the bristles in.

Topic 2 Exercises

1 Draw a diagram showing the inside structure of the Earth, and label the different parts.

2 Name the most abundant metal and the most abundant non-metal in the Earth's crust.

3 When lava flows out of a volcano, which part of the inside of the Earth has it come from? Why is it molten?

4 Why is the centre of the Earth solid, even though it is very hot?

5 How are sedimentary rocks formed? Why do they often contain fossils? What does this tell you about the ages of these rocks?

6 What type of rock is granite? How was it made? Name some of the different minerals that are found in granite. Give one use for the rock.

7 What does 'metamorphic' mean? How is this word used in geology?

8 How are stalactites and stalagmites formed?

9 What sort of water causes scale in a kettle? What chemical compound is the scale made from?

10 What mineral are rubies made of?

11 Describe, with the aid of a diagram, how the atmosphere changes as you go further from the Earth's surface.

12 Why must space rockets carry a supply of oxygen? Why is it always liquid oxygen?

13 List the main gases in the air. What percentage is there of each gas?

14 Describe some of the types of pollution that are found in the air.

15 Why might too much carbon dioxide in the atmosphere be dangerous?

16 What is a candle flame made of?

17 What methods can a fireman use to put out a fire? Will he use the same method for every fire?

18 Water flows out of a spring on a mountain side. Describe all the things that might happen to it before it gets back to the same place again.

19 Why is it important to treat sewage, before letting it flow into a river?

20 Describe what is done to river water, to make it safe to drink.

21 What is an aqueous solution?

22 What experiment would you carry out to see if your tap water contained dissolved solids?

23 What is meant by the word 'solubility'?

24 Look at the graph below. It shows the solubility curves for two different solutes in water. Study the graph and answer the questions that follow.

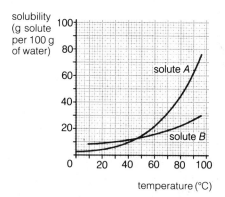

a) Which solute is more soluble in water at 20 °C?
b) Which solute is least soluble in water at 80 °C?
c) What is the solubility of solute A at 80 °C?
d) At what temperature does solute B have a solubility of 20 g per 100 g of water?
e) Why do the curves for the solute not go below 0 °C or above 100 °C?
f) A solution of solute A contains 66 g of A in 100 g of water at a temperature of 90 °C. It is cooled down to 30 °C. Use the graph to find out how much solute will crystallise out.
g) Estimate the solubility of solute B at 0 °C.
h) At what temperature is the solubility of both solutes the same?

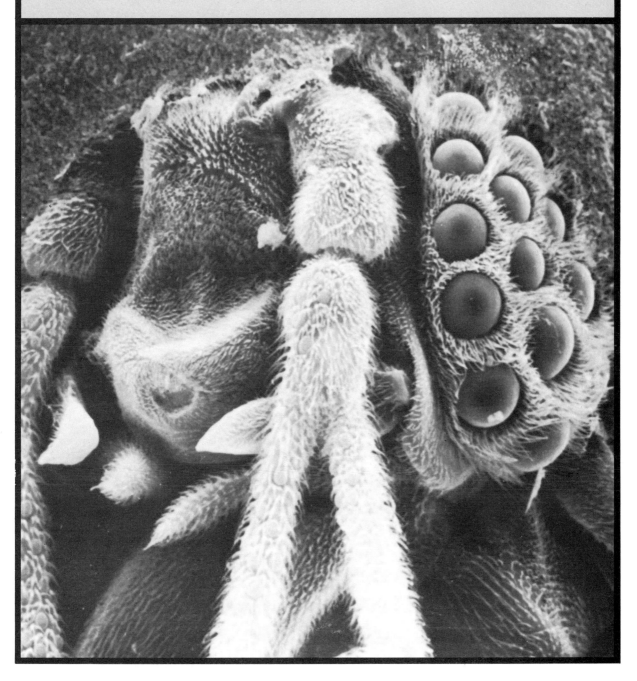

This is what a fly's head looks like under an electron microscope. A special version of the electron microscope has recently provided new evidence for the idea that everything is made of atoms.

3.1 Atoms

The word 'atom' is now part of our everyday language. But as scientists, you must know more precisely what it means.

What are atoms?

Atoms are tiny particles that everything is made of. 'Everything' includes you, me, the chair you are sitting on, and the air that you are breathing. All are made of atoms.

Atoms are so tiny that there are millions and millions of them in a single grain of sand. It is not surprising then that nobody has seen an atom. However, scientists are certain that they exist. Many of the things that happen in chemistry can only be explained if you believe in atoms, as you will see later.

The first people to think about atoms were the Ancient Greeks, more than 2000 years ago. A man called Democritus came up with the idea that if you keep cutting a bit of gold into tinier and tinier pieces, you will eventually come to a piece so small that it will be impossible to cut up any further. The tiny piece will be unsplittable. The old Greek word for unsplittable was 'atomos', and that is where we get the word *atom* from.

In modern times, the idea of the atom was taken up by a Quaker teacher and scientist called John Dalton. (See Figure 1.) In 1808, he published a set of rules about atoms:

1 Everything is made of atoms. Atoms cannot be split into anything smaller in a chemical reaction.
2 A substance made only of identical atoms is called an *element*.
3 There are many different kinds of atom, and that means there are many different elements.
4 Different atoms can join together during chemical reactions to make groups of atoms called *molecules*.

You can read more about elements and molecules later.

Of course, we now know that the first rule is wrong. Atoms *can* be split, in a *nuclear reactor* like the one shown in Figure 2. Inside the reactor, atoms split up giving out lots of heat. This heat is used to make steam to drive a turbine, which in turn drives a generator to make electricity. Nevertheless, the idea of an atom as a small, unsplittable particle is still very useful for describing a lot of what happens in chemistry.

The sizes of atoms

Atoms are very light. The usual unit of mass, the *gram*, is far too large to describe them. To get some idea of how light they are, think about this example:

   1 cm^3 of gold has a mass of about 19.3 g.
   It contains 147 000 000 000 000 000 000 000 atoms!

See Figure 3.

Figure 1   John Dalton believed atoms were unsplittable.

Figure 2   Atoms can be split inside a nuclear reactor to generate electricity. This is the Sizewell A nuclear power station. The turbine hall (note the steam) is in front of the reactor building.

Figure 3   1 cm^3 of gold contains 147 thousand million million million atoms.

Atoms are also very small. If you could put approximately 4 000 000 000 gold atoms end to end, they would stretch only 1 metre. See Figure 4.

For such tiny sizes, a new unit of length is used. It is called the *nanometre*. It has the symbol **nm**. One nanometre is one thousandth millionth of a metre, so an atom of gold is about 4 nm in diameter.

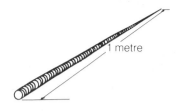

Figure 4 Atoms are extremely small. Four thousand million gold atoms would only measure 1 metre.

'Seeing' atoms

An atom of gold is far too small to be seen with an ordinary microscope. Indeed, no instrument has yet been made that will produce a clear picture of *any* atom. Some come fairly near to it however. For example, *X-rays* can be beamed through crystals so that they produce a pattern on a photographic film when they come out the other side. Such a pattern is called an X-ray diffraction pattern. The pattern is caused by the X-rays being scattered in different ways by the atoms in the crystal. Scientists can study the pattern and learn a great deal about the arrangement of the atoms in the crystal. An example is shown on page 44 (Figure 5).

The *field ion microscope* produces a similar pattern by pulling atoms, one by one, out of a metal and photographing them on a screen. An example is shown in Figure 5.

Biologists use an *electron microscope* to look at tiny parts of plant and animal cells. This instrument can 'see' groups of just a hundred or so atoms. See Figure 6.

In an American university, a special version of the electron microscope has taken a very blurred photograph of what is thought to be two uranium atoms, magnified five million times. See Figure 7.

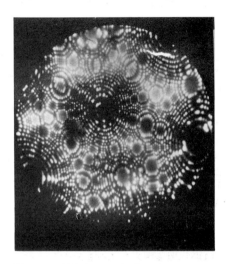

Figure 5 This is a typical pattern produced by a field ion microscope. The 'photo' is called a field ion micrograph.

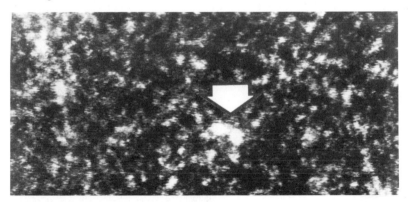

Figure 7 This photo is thought to show two uranium atoms (arrowed), magnified 5 million times.

Figure 6 This is what muscle cells look like under an electron microscope.

Exercises

1 Look around the room, and make a list of ten things in it that are made of atoms.

2 What did Democritus have to say about atoms?

3 If a substance is made of only one sort of atom, what is it called?

4 What are molecules, according to John Dalton?

5 What unit of length is used to measure the sizes of atoms?

6 About how many gold atoms would fit side-by-side on a line 1 cm long?

7 Has anyone ever seen an atom?

3.2 Arranging atoms

There are many different kinds of atoms and they can be joined together in different groups and different ways.

The different kinds of atoms

Over 100 different kinds of atoms have been found on the Earth, or made artificially in laboratories. That means there are over 100 elements. Remember, an *element* is a pure substance that contains only one sort of atom.

Each element has a *symbol* and the complete list of the elements and their symbols is called the *Periodic Table*. There is a Periodic Table on page 70, and you can read more about elements and their symbols on page 39.

Different molecules

Some atoms can join together in small numbers, to make groups called *molecules*. Sometimes a molecule is made up of two or more atoms of the same sort. Examples are the molecules of oxygen, hydrogen, and nitrogen, shown in Figure 1. But more often, molecules are made of atoms of different sorts, like the molecules of carbon dioxide, water, and ammonia shown in Figure 2.

Three states of matter

Everything in the world is either a *solid*, a *liquid*, or a *gas*. See Figure 3. Solid, liquid, and gas are known as the three *states of matter*.

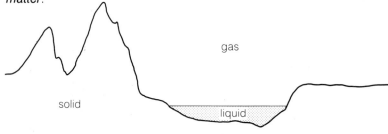

Figure 3   Solid, liquid, and gas are the three states of matter.

Solids, liquids, and gases are quite different from each other. Solids are hard and difficult to break. Think of stone and iron. Liquids are runny and take the shape of their containers – examples are water and milk. Gases can be squashed or *compressed* easily, like air in a bicycle tyre. We say that solids, liquids, and gases have different *properties*. The differences are due to the ways their atoms and molecules are arranged.

Solids

Here is a description of a solid such as iron:

1   It is strong, but can be bent or shaped with difficulty. (See Figure 4.)

2   It keeps its shape, but if it is heated it melts and changes into a liquid.

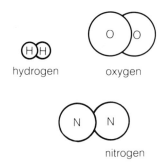

Figure 1   These molecules are made of the same atoms.

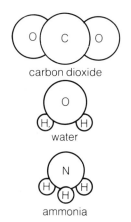

Figure 2   These molecules are made of different atoms.

Figure 4   A solid such as iron is difficult to bend but is very strong. The first iron bridge in England was built at Ironbridge in Shropshire.

These properties can be explained by the arrangement of the atoms in the solid. Look at Figure 5A. The atoms are all packed closely together in a regular pattern. They vibrate a little, but they cannot move around unless a whole layer moves at the same time. (This is what happens when iron is hammered or bent.) When the solid is heated, the atoms receive energy which makes them vibrate more and more. So eventually the regular arrangement breaks down. At this point, the solid has melted. See Figure 5B.

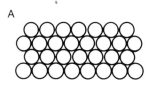

Figure 6 A liquid takes the shape of its container.

Figure 5 **A:** Atoms in a solid are arranged in a regular way. **B:** This regular arrangement breaks down if the solid is heated till it melts.

Liquids

Here is a description of a liquid such as water:

1 It has no definite shape – it takes the shape of its container. (See Figure 6.)

2 It has no 'strength' but it can't be squashed.

3 When it is heated, it boils and turns into a gas.

In a liquid like water, the molecules are close together, like the atoms in iron. But they are *not* arranged in a regular way. They can slide over each other easily, which means the liquid can take up any shape. However, since the molecules are already close together, they cannot be pushed closer without a great deal of effort. Machines like JCB's rely on this property of liquids. Their moving parts are operated by oil under pressure in pipes and pistons. See Figure 7.

When a liquid is heated, the molecules obtain more and more energy. They move around faster and faster until they knock each other so far apart that the liquid changes into a gas. Look at Figure 8.

Figure 7 The moving parts of a JCB are operated by oil under pressure.

Figure 8 When a liquid is heated, the molecules obtain more and more energy, till the liquid becomes a gas.

Gases

Here is a description of a gas such as oxygen:

1 It has no definite shape – it will fill any container that holds it.

2 It can easily be compressed to hundreds of times atmospheric pressure in steel cylinders. (See Figure 9.)

3 When it is heated, it just gets hotter and hotter.

In a container of oxygen, the molecules are much further apart than the molecules in water. That means there is lots of room for them to be pushed closer together – the gas can be compressed. In a gas, the molecules move around very quickly. The hotter the gas gets, the quicker they move. A gas can't change into anything else no

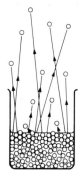

Figure 9 A gas can be compressed to several hundred times atmospheric pressure.

matter how hot it gets. But if it is cooled, the molecules slow down and get closer and closer together, until eventually a liquid is formed. Look at Figure 10.

It's all only an idea

You have just seen the way that scientists explain the behaviour of solids, liquids, and gases. All this is only an idea or 'theory' because no-one has ever seen a molecule. However, the description does work. The word 'kinetic' means movement, so this way of describing the movement and arrangement of atoms and molecules is called the *Kinetic Theory*.

Figure 10 When a gas cools, it changes back to a liquid.

Exercises

1 Name one molecule made of atoms of the same type.

2 Name one molecule made of atoms of different types.

3 How many different types of atoms are there in an element?

4 Write down two ways in which a solid differs from a liquid.

5 Name the three states of matter.

6 In which of the three states of matter do molecules move fastest? In which are the molecules closest together? In which are the molecules close together but *not* arranged in a regular pattern?

7 Look at the pictures below. Make a list of the changes of state that are going on. In other words, where can you see a solid changing into a liquid, etc?

Melting ice-cream

Water vapour droplets visible as a rocket is launched

Molten metal

Solid carbon dioxide being used to produce fog on stage

3.3 Elements and compounds

Hydrogen and oxygen are elements, but water, which is made from these two gases, is a compound.

What are elements?

An element is a pure substance that cannot be made into anything chemically simpler. Some examples of elements are carbon, silver, sulphur, gold, copper, chlorine, and iron. The list could go on for many lines because there are 92 different elements found on Earth, and about 15 radioactive ones that have been man-made in nuclear reactors. You can see a complete list of elements on page 70.

Elements can also be explained in terms of atoms. If you could see into a copper pipe you would find that all the copper atoms were identical. (See Figure 1.) Similarly, if you could look into a piece of sulphur, you would find that all the sulphur atoms were identical. But the copper atoms would be different in size from the sulphur atoms.

Elements are pure substances whose atoms are all identical. The atoms of different elements are different in size.

Symbols

Every element has its own symbol, which is the short way to write it. The symbol is usually the first letter of the element's name, or the first letter plus another letter. The symbol always begins with a capital letter. You will soon remember many of the symbols without having to think about them. For example, look at the symbols for the elements in Figure 2. They are very easy to learn.

Not all the elements were discovered at the same time. Some, like gold, iron, silver, tin, and lead have been known for thousands of years and have names from Greek and Latin. For example the name 'hydrogen' is made from the Greek phrase 'hydro gennao' which means 'I generate water' because this is what hydrogen does when it burns. Similarly, oxygen comes from 'oxus gennao' which means 'I generate acid' because many substances form acidic compounds when they burn in oxygen. Some symbols that come from Greek and Latin names are shown in Figure 3.

Some elements are named after the countries in which they were first discovered, or even the universities in which they were first made. Look at Figure 4 for examples.

Figure 1 Copper is an element. All the atoms in these pipes are identical.

name	symbol
hydrogen	H
nitrogen	N
iodine	I
oxygen	O
sulphur	S
carbon	C

Figure 2 The symbol for these elements is simply the first letter of the element's name.

name	symbol and derivation
sodium	Na from the Latin *natrium*
lead	Pb from the Latin *plumbum*
mercury	Hg from the Latin *hydrargyrum*
gold	Au from the Greek *aurum*

Figure 3 The symbol for these elements comes from a foreign language.

name	symbol and derivation
strontium	Sr after Strontia in Scotland
magnesium	Mg after Magnesia in Asia Minor
americium	Am after America
californium	Cf after California

Figure 4 These elements are named after places.

Some elements are named after mythological characters or planets in space (Figure 5), and some are even named after the Latin names for colours (Figure 6).

name	symbol and derivation
titanium	Ti after the Greek god Titan
cobalt	Co after a German sprite
thorium	Th after the Scandinavian god Thor

Figure 5 These elements are named after mythological characters.

name	symbol	colour
rubidium	Rb	red
indium	In	indigo
caesium	Cs	blue-grey

Figure 6 The symbol for these elements comes from the Latin word for a colour.

What are compounds?

Only 92 different elements occur naturally on Earth. That means only 92 different kinds of atoms occur naturally. However, different atoms can join together in groups of two or more, making hundreds of thousands of different molecules. Substances with molecules that contain different atoms are called *compounds*.

For example, sulphur is an element. When yellow sulphur powder is heated in air, it melts and burns with a dark-blue flame. The sulphur slowly disappears and a choking, invisible gas called sulphur dioxide is formed in its place. The sulphur has combined chemically with oxygen from the air. Figure 7 shows what happens to the atoms and molecules. During the chemical reaction, the sulphur atoms and oxygen molecules rearrange themselves to make molecules of sulphur dioxide.

Sulphur dioxide is *not* an element, since its molecules contain two different kinds of atoms. It is a compound. *The molecules of a compound contain different kinds of atoms*.

Here is another example. Hydrogen and oxygen are both invisible gases. If they are mixed in a balloon, nothing happens. The balloon simply contains a mixture of hydrogen molecules and oxygen molecules. But if someone holds a match to the balloon, there is a loud bang! The two elements have combined explosively to form a new substance, water. The water will appear as condensation on the inside of the burst balloon. Look at Figure 8. Water is a compound because each water molecule contains two hydrogen atoms joined to one oxygen molecule.

Mixtures and compounds

Not everything that contains atoms of different elements is a compound. In Topic 1, you learned about mixtures. Some examples of mixtures are:

solutions (solids + liquids) – for example, sea water
suspensions (solids + liquids) – for example, muddy river water
emulsions (liquids + liquids) – for example, salad dressing.

These are not compounds, because the atoms of the separate substances are not *chemically combined* to form a new substance. Sea water can be evaporated to obtain the salt, the dirt can be filtered from river water, and salad dressing separates into oil and

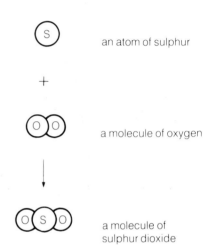

an atom of sulphur

+

a molecule of oxygen

a molecule of sulphur dioxide

Figure 7 When sulphur combines with oxygen the compound sulphur dioxide is formed.

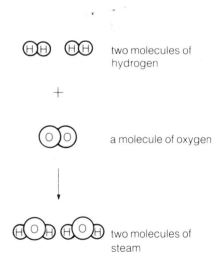

two molecules of hydrogen

+

a molecule of oxygen

two molecules of steam

Figure 8 When hydrogen combines explosively with oxygen the compound water is formed.

vinegar on standing. But compounds can only be separated into their elements by further chemical reactions.

Exercises

1 What is an element?

2 Write down the names of four elements.

3 What can you say about the sizes of all the atoms in a piece of carbon?

4 Use the Periodic Table on page 70 to find out the symbols for the following elements:

sodium   potassium   phosphorus
lead   tin   iron.

5 Using the Periodic Table, find out which elements have the following symbols:

Si   Ar   Co
Cu   W   Pt

6 Some elements have been named after famous scientists. Find out which elements are represented by the following symbols. Then find out which scientists they are named after, and what these scientists were famous for.

Es   No   Md

7 What is a compound?

8 If you could see the molecules inside a compound, what would you notice about the atoms in them?

9 The photo below shows the salt that is left when sea water evaporates. What does this tell you about the atoms in sea water?

10 Water is a compound made from oxygen and hydrogen. Air is a mixture of oxygen and nitrogen. How would you explain to someone the difference between a mixture and a compound, using water and air as examples?

41

3.4 Evidence for atoms and molecules

It is easy to explain what happens in several simple experiments, if you believe in atoms and molecules. But without them, things become very difficult indeed.

Dissolving

When a sugar lump dissolves in water, it disappears and leaves a clear liquid. The sugar must still be there, because you can taste it, but why can't you see it?

The question is easily answered if you think of sugar as being made of millions of molecules all packed closely together, as in Figure 1. Water is also made of molecules, but these are moving around freely, as shown in Figure 2. When sugar dissolves in water, the sugar molecules simply move into the spaces between the water molecules. The sugar and water molecules mix together and make a solution. A solution is just a mixture of molecules. Look at Figure 3.

Diffusing

When a gas escapes into the air, it quickly spreads out and 'disappears'. Just like sugar and water, a gas is made of molecules. So is the air. When the gas is released into the air, all the molecules get mixed up. The gas hasn't really disappeared. It has spread out and mixed so thinly with the air that it can no longer be easily detected.

The mixing of gases like this is called *diffusion*. Figure 4 shows how a car's exhaust fumes diffuse into the air.

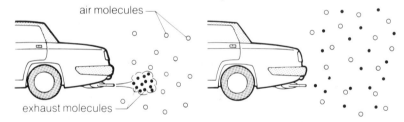

air molecules
exhaust molecules

Figure 4 A car's exhaust molecules soon diffuse into the air.

There is also the tale of Nelson's last breath. When Admiral Nelson gasped his last breath on the deck of HMS *Victory*, the molecules he breathed out diffused into the air and have been diffusing ever since. Someone has calculated that on average, every time you take a breath, you take in one of those molecules!

Diffusion takes place in liquids too. If you squirt coloured bubble bath into the water at one end of your bath, the colour gradually spreads out and mixes through all the water. The molecules of the coloured liquid mix with the molecules of bath water in the same way that exhaust fumes mix with the air and sugar molecules mix with water. These are *all* examples of diffusion. Of course, if you

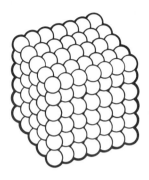

Figure 1 A sugar cube is made of millions of molecules packed closely together.

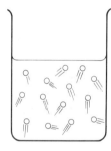

Figure 2 The molecules in water are not as closely packed as those in sugar. They can move quite freely.

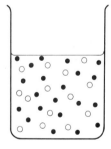

Figure 3 A solution is a mixture of molecules. A sugar solution is simply a mixture of sugar molecules and water molecules.

waited for diffusion to mix your bubble bath for you, your bath water would get cold. This is because diffusion in liquids is a slow process. Diffusion in gases is much quicker. Think how fast a smell can travel.

Crystallising

When a solution of sugar in water is left in a warm room, the water evaporates into the air, leaving sugar crystals behind. This process is called *crystallisation*. When the water molecules leave the solution, the sugar molecules are left arranged regularly in rows and layers. Look at Figure 5. *Crystals* have flat surfaces and straight edges and regular shapes because their molecules are always arranged in a regular way. You can read more about crystals on the next page.

Figure 5 Crystals have flat surfaces, straight edges, and regular shapes. These are sugar crystals, magnified 1½ times.

Brownian motion

In 1927 Robert Brown, a botanist, made an interesting discovery. He was looking through a microscope at pollen grains floating on water, and he noticed that the grains were dancing around in a very jerky way. He could only explain this in terms of molecules. The water molecules were bumping into the pollen grains, and making them move around. The water molecules were themselves far too small to be seen, but their effect on the pollen was clear.

The movement of the pollen became known as *Brownian motion*. Brownian motion can be seen in the laboratory using a smoke cell. Look at Figure 6. When the tiny particles of soot in smoke are magnified under a microscope, they can be seen jostling around. This jostling is caused by air molecules bumping into the soot particles.

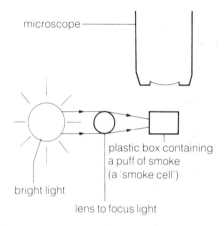

Figure 6 A smoke cell for seeing Brownian motion.

Exercises

1   Describe in your own words what happens to the molecules when: **a)** salt dissolves in water **b)** a pool of salt water dries up in the sun.

2   Look at the diagram on the right. Jar *A* contains a red gas. Jar *B* contains air. What will happen when the glass plate between them is pulled away? What name is given to this process?

3   What did Robert Brown see in his microscope?

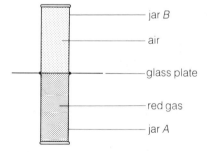

3.5 Crystals

Everyone knows that crystals have flat surfaces, straight edges and regular shapes, but you need to think about atoms and molecules to understand why they are like that.

Shapes

Crystallographers are people who study crystals. They investigate the internal structure of the crystals and they put them into *systems* or *habits* according to their external shape. For example, a very common substance is sodium chloride (salt). The atoms of sodium and chlorine inside the crystal are arranged in a *cubic lattice* and the outside shape of the crystal is that of a cube. Look at Figure 1. Pyrite ('fool's gold') has a cubic shape too. Look at Figure 2.

Figure 3 An alum crystal is octahedral in shape. It has 8 even surfaces.

Figure 1 Crystals of salt are cubic shaped.

Figure 2 Crystals of pyrite are cubic shaped.

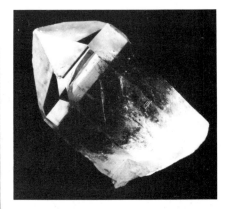

Figure 4 A quartz crystal is hexagonal in shape.

However, another crystal with a cubic internal structure, alum, forms crystals which are *octahedral* in shape. See Figure 3. Diamond behaves in this way too. Quartz, the shiny bit in granite, is another shape. Its crystals are *hexagonal*. Look at Figure 4.

Seeing inside crystals

Scientists can study the arrangement of atoms inside a crystal using *X-rays*. The rays are beamed through a powder or a single small crystal, and photographed when they come out the other side. This process is called *X-ray diffraction*. The pattern produced by the X-rays on the film can be used to work out the arrangement of the atoms. Figure 5 shows a crystal being fitted to an X-ray diffractometer and the pattern that X-rays produce for it. This pattern doesn't itself show the arrangement of the atoms, however. The scientist must do a lot of sums to work out how the X-rays will have been scattered or *diffracted* by the atoms, as they pass through the crystal.

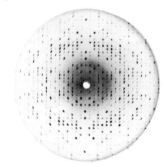

Figure 5 The photo above shows a crystal being fitted to an X-ray diffractometer. The photo below shows the pattern produced on the film by the X-rays.

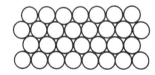

Figure 6 The photo on the left shows a crystal of calcite. You can see the cleavage lines where the crystal would break between layers of atoms. The photo on the right shows layers of mica being peeled from a crystal with a scalpel.

Breaking crystals

If a crystal is hit with a hammer it will probably shatter into many pieces. However, some crystals break in an even way. The forces holding the atoms together in the crystal are stronger in some directions than others. If it is hit in the right direction, the crystal may split between lines and layers of atoms. This is called *cleavage*. Figure 6 shows a crystal of *calcite*. You can see some cleavage lines which show where the crystal would break in a very clean and even way. *Mica* is another substance with crystals that can be cleaved in this way. You can peel off thousands of layers of atoms at a time with your fingers or a scalpel. See Figure 6.

Figure 7 Kimberley Great Hole in South Africa is the remains of a diamond mine.

From rough crystals to diamond rings

Diamonds are found in parts of the Earth's crust that have been subjected to intense pressure and heat. One such place is the Kimberley Great Hole in South Africa (see Figure 7). It is 300 m deep and before it closed in 1909, about 3 tonnes of diamonds had been found there. When the rough diamond is shaped, a skilled man cleaves it with a sharp steel blade so that it splits evenly between the layers of atoms. Next, the crystal is cut to shape with a tiny diamond-edged circular saw. Finally it is polished with diamond powder until the surfaces of the crystal are smooth. These surfaces, or facets, are cut so that the diamond reflects light back at you to give that familiar sparkle. Despite the fact that diamond is the hardest substance, it can be cut because its atoms are arranged in a regular way.

Figure 8 This photo shows the crystals in a fracture surface of zinc, magnified 8 times.

Metals

Metals have crystals too. Look at Figure 8. The atoms in metals are arranged in a very regular and even way. Look at Figure 9. The atoms in this metal are *hexagonally close packed*. It is this regular arrangement that gives a metal strength, but also allows it to be shaped by hammering and stretching. You can read more about the structure of metals in Topic 5.

Figure 9 This arrangement of atoms in a metal is called hexagonal close packing.

Exercises

1 Look at some sugar crystals with a hand lens. What shape are they?

2 Metals have crystals. So why is it that metal objects can come in all shapes and sizes?

3 Diamond is the hardest naturally-occurring crystal, and yet it can be cut into intricate shapes. Describe the ways in which this is done.

3.6 Understanding formulae

A compound has a formula, just as an element has a symbol.

What a formula shows

A compound is made up of different elements. Its *formula* is made up from the symbols of these elements. Figure 1 shows a molecule of water, and the formula for water. The '2' after the hydrogen in the formula shows that there are two atoms of hydrogen in a molecule. There is only one atom of oxygen, but the '1' is never written in the formula. Look again, and compare the picture of the molecule with what the formula tells you.

Figure 2 shows another example – a molecule of carbon dioxide, and the formula for carbon dioxide. The formula shows that there are two atoms of oxygen joined to one atom of carbon. Figure 3 shows a molecule of ammonia and the formula for ammonia. The '3' shows that there are three atoms of hydrogen joined to one atom of nitrogen.

Sometimes, brackets are used in a formula. Look at these two compounds:

$Cu(OH)_2$ copper hydroxide $Al_2(SO_4)_3$ aluminium sulphate

In copper hydroxide, the brackets around the OH mean that there are two oxygen atoms and two hydrogen atoms joined to the one copper atom. Just as in mathematics, everything inside the brackets is multiplied by two. Similarly, in aluminium sulphate, the 2 after the aluminium shows that there are two aluminium atoms, and the 3 outside the brackets means three times everything inside. In other words, three sulphur atoms and twelve oxygen atoms.

The plural of 'formula' is *formulae*. Try the exercises at the bottom of the next page to make sure you really understand how to read formulae.

-ide compounds

Look at the compounds and their formulae in Figure 4. You should notice two things about them:

1   Each compound contains just *two different elements*.
2   Their names all end in *-ide*.

When a metal element combines with another element to make a compound, the name of the metal element stays the same, and the name of the second element changes its ending to '-ide'.

So   oxygen changes to oxide
     chlorine changes to chloride
     sulphur changes to sulphide, and so on.

Don't be confused when there are numbers in the formula. These just tell the *numbers of atoms* of each element in the molecule. The rules for making the *name* remain just the same. Look at Figure 5 for some examples.

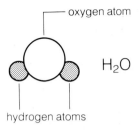

Figure 1   A water molecule has two hydrogen atoms and one oxygen atom.

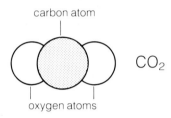

Figure 2   A carbon dioxide molecule has two oxygen atoms and one carbon atom.

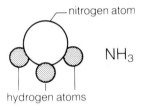

Figure 3   An ammonia molecule has three hydrogen atoms and one nitrogen atom.

calcium oxide	CaO
iron sulphide	FeS
sodium chloride	NaCl
silver bromide	AgBr
potassium iodide	KI

Figure 4   When a metal combines with a non-metal to form a compound, the name of the metal stays the same. The name of the non-metal changes its ending to -ide.

lithium oxide	Li_2O
potassium sulphide	K_2S
iron chloride	$FeCl_3$

Figure 5   The name of the compound is not affected by any numbers in the formula.

-ate compounds

Look at the compounds and their formulae in Figure 6. You should notice two things about them:

1 Each compound contains *oxygen atoms*.

2 Their names all end in *-ate*.

SO_4 is called the *sulphate group*. It contains one sulphur atom and four oxygen atoms.

NO_3 is called the *nitrate group*. It contains one nitrogen atom and three oxygen atoms.

CO_3 is called the *carbonate group*. It contains one carbon atom and three oxygen atoms.

PO_4 is called the *phosphate group*. It contains one phosphorus atom and four oxygen atoms.

Remember not to be confused by brackets or numbers. They do not alter the name of the compound. Look at the examples in Figure 7.

copper sulphate	$CuSO_4$
potassium nitrate	KNO_3
calcium carbonate	$CaCO_3$
calcium phosphate	$Ca_3(PO_4)_2$

Figure 6 These compounds all contain oxygen. The name of each compound ends in -ate.

sodium sulphate	Na_2SO_4
iron nitrate	$Fe(NO_3)_2$
silver carbonate	Ag_2CO_3

Figure 7 The name of the compound is not affected by any numbers in the formula.

Two more groups

NH_4 is the formula for the *ammonium group*. This contains one nitrogen atom and four hydrogen atoms. Two examples of compounds containing the ammonium group are:

ammonium chloride, which has the formula NH_4Cl
ammonium sulphate, which has the formula $(NH_4)_2SO_4$.

Don't confuse the ammonium group with ammonia gas, which has the formula NH_3.

OH is the formula for the *hydroxide group*. This contains one oxygen atom and one hydrogen atom. Two compounds which contain the hydroxide group are:

sodium hydroxide, which has the formula NaOH
copper hydroxide, which has the formula $Cu(OH)_2$.

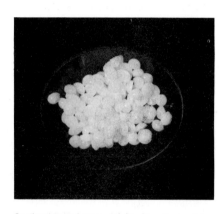

Sodium hydroxide contains the hydroxide group. Its formula is NaOH.

Exercises

1 a) Write down the names of the elements contained in each of the following compounds. (Remember, there is a Periodic Table at the back of the book to help you.)
b) How many atoms of each element are there in one molecule of the compound?

magnesium bromide	$MgBr_2$
zinc chloride	$ZnCl_2$
aluminium hydroxide	$Al(OH)_3$
ammonium sulphate	$(NH_4)_2SO_4$
calcium nitrate	$Ca(NO_3)_2$

2 Try to write down the formulae of these compounds. They are all simple ones that do not involve brackets.

a) barium oxide   b) nickel sulphide
c) lithium chloride   d) potassium bromide
e) silver iodide   f) ammonium hydroxide

3 Write down the names of the compounds represented by these formulae:

a) Cs_2SO_4 b) $Sr(NO_3)_2$ c) Hg_2SO_4
d) $(NH_4)_2CO_3$ e) $Mg(OH)_2$

3.7 Writing formulae

You have learned to put names to formulae. Now you can learn the rules for writing formulae from names.

Combining numbers

When you study chemistry later on in the school, you will learn that the way atoms combine together in a compound depends on the structure of the atoms. For the present, it is enough to know the *combining numbers* of the elements. Using these numbers, you can tell how many atoms of different elements will combine together. And that means you can tell the formulae of their compounds.

The combining number of an element is sometimes called its *valency*. Figure 1 shows the combining numbers or valencies for many of the elements or groups you will meet.

Rules for writing formulae

When you work out a formula from a name, you must follow these rules:

1 Write down the name of the compound.

2 Write down the symbols for the elements and groups in the compound.

3 Write down their combining numbers.

4 Adjust the numbers of atoms of each element or group until the + combining numbers equal the − combining numbers.

Look at these examples.

Example 1

1	sodium	chloride	This is the name of the compound.
2	Na	Cl	These are the symbols of the elements.
3	+1	−1	These are the combining numbers.

4 Now, do the + combining numbers equal the − combining numbers? Yes, they do. *The formula of sodium chloride is NaCl.*

Example 2

1	copper	bromide	This is the name.
2	Cu	Br	These are the symbols.
3	+2	−1	These are the combining numbers.

4 The + combining numbers and the − combining numbers are not equal. You need to add another bromide. This gives:

| Cu | Br | Br |
| +2 | −1 | −1 |

Now the numbers are equal. *The formula is written as $CuBr_2$.* The 2 shows that there are two bromides for every copper. Note that the 2 is written a little below the line.

element or group	combining number
H, hydrogen	+1
Na, sodium	+1
K, potassium	+1
Ag, silver	+1
NH_4, ammonium	+1
Ca, calcium	+2
Mg, magnesium	+2
Cu, copper	+2
Zn, zinc	+2
Pb, lead	+2
Fe, iron (*)	+2
Al, aluminium	+3
Fe, iron (*)	+3
OH, hydroxide	−1
NO_3, nitrate	−1
Cl, chloride	−1
Br, bromide	−1
I, iodide	−1
SO_4, sulphate	−2
CO_3, carbonate	−2
O, oxide	−2
S, sulphide	−2
PO_4, phosphate	−3

*Note that iron has two combining numbers. You will see both being used, on the next page.

Figure 1 The combining numbers of some elements and groups.

Example 3

1 iron(II)   hydroxide

2 Fe   OH

3 +2   −1

The (II) in the name shows that the combining number of iron in this compound is +2.

4 The + numbers do not equal the − numbers. You need two hydroxide groups for each iron atom. This gives:

Fe   OH   OH
+2   −1   −1

Now the numbers are equal. *The formula of iron(II) hydroxide is Fe(OH)₂*. Brackets are needed because the 2 refers to both the O and the H.

Example 4

1 iron(III)   sulphate

2 Fe SO_4

3 +3   −2

The (III) in the name shows that the combining number of iron in the compound is +3.

4 The + numbers do not equal the − numbers. You need two iron atoms and three sulphate groups. Like this:

Fe   Fe SO_4 SO_4 SO_4
+3   +3   −2   −2   −2

Now the numbers are equal. *The formula of iron(III) sulphate is Fe₂(SO₄)₃*. The 2 shows that there are two iron atoms. The 3 shows that there are three sulphate groups.

These examples should help you to understand the rules for working out the formulae of compounds. Now try the exercises that follow.

The chemical name for salt is sodium chloride. Its formula is NaCl.

The active ingredient in oven cleaner is sodium hydroxide. Its formula is NaOH.

Exercises

1 Write down the names of:
   a) 3 elements that have a combining number of +1
   b) 3 elements that have a combining number of +2
   c) 2 elements that have a combining number of −1
   d) 2 groups that have a combining number of −2.

2 Work out the formulae of these compounds:
   a) lithium hydroxide   b) calcium iodide
   c) silver nitrate   d) sodium sulphate
   e) aluminium oxide   f) potassium sulphide
   g) iron(II) carbonate   h) iron(III) chloride
   i) ammonium phosphate   j) lead nitrate
   k) magnesium carbonate   l) calcium hydroxide
   m) copper oxide   n) aluminium iodide

3.8 Chemical reactions

Whenever elements combine to form compounds, or compounds split up and form new compounds, a chemical reaction takes place.

What is a chemical reaction?

On page 40 you saw that atoms of sulphur react with molecules of oxygen, on heating, to form molecules of a gas called sulphur dioxide. Look at Figure 1. The reaction is a *chemical reaction*, because a new substance is formed. It can be described in words and in symbols, like this:

sulphur	+	oxygen	→	sulphur dioxide
S	+	O_2	→	SO_2

Sulphur and oxygen are called the *reactants*.	The arrow shows that a reaction is taking place.	Sulphur dioxide is called the *product*.

These short ways of describing a reaction are called *equations*.

Recognising chemical reactions

All chemical reactions have four things in common:

1 The product or products of the reaction are *new substances* with new names and formulae.
2 The products usually look quite different from the reactants.
3 The chemical reaction involves an *energy change* of some sort.
4 A chemical reaction is usually quite difficult, if not impossible, to reverse.

Examples of chemical reactions

The four common properties of chemical reactions can be seen in the following examples.

Lighting a Bunsen burner For this reaction methane gas (North Sea gas) is mixed with air (which contains oxygen). When the methane and oxygen react, they form steam and carbon dioxide.

This chemical reaction is called *combustion*. It is illustrated in Figure 2. The products are completely different from the reactants. Energy is given out in the form of heat and light. The steam and carbon dioxide cannot be turned back into methane and oxygen.

This is the equation for the reaction:

$$CH_4 \ + \ 2O_2 \ \rightarrow \ CO_2 \ + \ 2H_2O \ + \ \text{heat and light}$$

methane   oxygen   carbon dioxide   steam

Setting off a firework Look at Figure 3. The reactant in this case is gunpowder. The products are ash, hot gases, and sparks. A lot of energy is released in the form of heat, light, and sound. Do you think you could turn the products back into a firework again?

In each example above, a small amount of energy is needed to start the chemical reaction off. This is usually provided by a burning

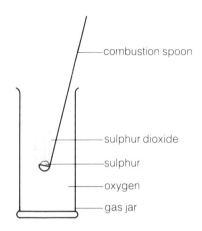

Figure 1 When sulphur is heated in oxygen a chemical reaction takes place. The new substance formed is sulphur dioxide.

Labels: combustion spoon, sulphur dioxide, sulphur, oxygen, gas jar

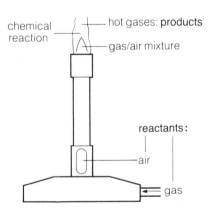

Figure 2 Combustion is a chemical reaction.

Labels: chemical reaction, hot gases: products, gas/air mixture, reactants:, air, gas

Figure 3 When a firework burns a chemical reaction takes place.

match. But once the reaction is under way, far more energy is given out than was put in at the start. Reactions like these, that give out heat energy, are called *exothermic reactions*.

Making iron sulphide   Your teacher might let you do this experiment for yourself. When yellow sulphur powder is mixed with grey iron filings, and the mixture is heated, it glows and bubbles. When cool, it sets into a brittle black solid. This solid is called iron sulphide.

The product looks quite different from the chemicals you start with. It has a new name and formula. Heat energy is given out during the reaction. And it would be extremely difficult to turn iron sulphide back into iron filings and sulphur powder again. A chemical reaction has certainly taken place. The equation for it is:

$$Fe \quad + \quad S \quad \rightarrow \quad FeS$$
   iron   sulphur   iron sulphide

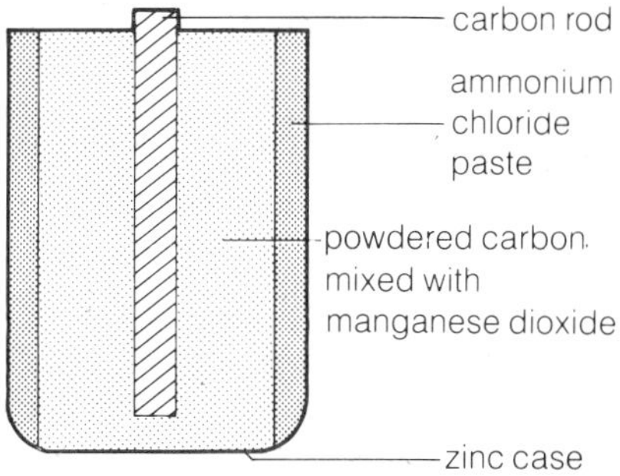

Using a torch battery   This involves a chemical reaction that gives out energy in the form of electricity. The chemicals are inside the battery case, as shown in Figure 4. When the two ends of the battery are connected by switching on the torch, the chemicals react together and electricity is produced. It flows from the battery to the bulb. Inside the bulb, the electrical energy is changed into heat and light energy.

Changes that are not chemical reactions

Not everything that happens in a test tube or beaker is a chemical reaction. Here are some examples of changes that are *not* chemical reactions.

Dissolving sugar in water   This change can be shown as:

sugar + water → sugar solution

The product looks different from the reactants, and energy in the form of heat may be needed to make the sugar dissolve. But a new substance has not been formed. The sugar is still there in the water. You could get it back by evaporating the water.

Boiling water   Heat energy is needed to change water to steam. Steam looks quite different from water. (You can't see true steam.) But both water and steam are made up of water molecules, so no new chemical has been formed. The steam can easily be changed back to water again, by condensation.

All changes like dissolving, melting, boiling, freezing, condensing, and crystallising are called *physical changes*. They are not chemical reactions. See Figure 5.

carbon rod
ammonium chloride paste
powdered carbon. mixed with manganese dioxide
zinc case

Figure 4   When a torch is switched on, a chemical reaction takes place inside the battery case. The chemicals involved are shown in this drawing.

Figure 5   Molten iron looks different from solid iron. But it is not a new substance. Melting is not a chemical change.

Exercises

1   Think about each change in the list below. Decide whether it is a chemical reaction or a physical change, and give reasons for your answer:
   a) baking a cake   b) painting a piece of wood
   c) burning coal   d) adding sugar to tea
   e) heating a piece of iron to 2000 °C and then letting it cool down again
   f) burning a candle   g) driving a car.

2   What is meant by an exothermic reaction? Give one example.

3   Describe a chemical reaction that produces energy in the form of:
   a) heat   b) light   c) sound
   d) electricity   e) movement
   f) heat, light, sound, and movement.

3.9 Writing chemical equations

Equations are chemical sentences. To write them correctly, you must use the right grammar.

Rules for writing chemical equations

As you saw on page 50, a chemical reaction can be described by an equation. The equation can be written using either words or symbols. Symbols are easier to look at, and they tell you more about how the reactants and products are made up.

To write an equation with symbols, you must first know the symbols, or where to look them up. Next you must be able to work out the formulae of compounds, as explained in Unit 3.7. You cannot write equations if you are not able to do this.

Then you must follow these rules:
1 Write down the reaction in words.
2 Write down the symbols and formulae for each of the substances in the reaction, and make sure they are correct.
3 Make the number of atoms of each element on both sides of the equation the same.

Examples of chemical equations

Carbon burning in air   The carbon reacts with the oxygen in air.

1   carbon +   oxygen   → carbon dioxide
2   C   + O_2 → CO_2
             oxygen gas is   this is the
             always made   correct formula
             of molecules

3   Now count the different atoms on each side of the equation. There is one carbon atom and two oxygen atoms on each side. So there is no problem. The equation is *balanced*:

$$C + O_2 \rightarrow CO_2$$

Heating calcium carbonate   When this solid is heated strongly, it decomposes to form carbon dioxide and calcium oxide.

1   calcium carbonate   →   carbon dioxide   +   calcium oxide
2      $CaCO_3$ → CO_2 +   CaO
     check that this   you know this   check that this
     is correct   one already   is correct

3   Now count the different atoms on each side of the equation. There is one calcium atom and one carbon atom on each side, and there are three oxygen atoms on each side. Again there is no problem – the equation is balanced:

$$CaCO_3 \rightarrow CO_2 + CaO$$

Dissolving zinc in hydrochloric acid   Zinc reacts with this acid to form a gas called hydrogen and a solution of a compound called zinc chloride.

1   zinc +   hydrochloric acid   →   hydrogen   +   zinc chloride
2   Zn +   HCl   → H_2 $ZnCl_2$
           learn this formula   hydrogen gas   check that
                                     is always made   this is
                                     of molecules   correct

52

3 Now count the atoms. Zinc is all right – there is one zinc atom on each side of the equation. However there are two hydrogen atoms and two chlorine atoms on the right-hand side, but only one of each kind on the left. The equation is not balanced.

You can solve the problem by putting a 2 *in front of* the HCl on the left-hand side. The equation becomes:

$Zn + 2HCl \rightarrow H_2 + ZnCl_2$

The 2 refers to everything in the formula that follows it. Note that it would be wrong to write H_2Cl_2 instead, because that would change the formula of hydrochloric acid. The equation is now balanced:

$Zn + 2HCl \rightarrow H_2 + ZnCl_2$

Reacting sodium hydroxide with sulphuric acid In this reaction, water and a compound called sodium sulphate are formed.

1 sodium hydroxide + sulphuric acid → sodium sulphate + water

2 $\quad$ NaOH $\quad$ + $\quad$ H_2SO_4 $\quad$ → $\quad$ Na_2SO_4 $\quad$ + H_2O
$\quad$ check this $\qquad\qquad$ learn this one $\qquad\quad$ check this

3 Now count the atoms. The equation is not balanced, because the numbers of sodium and hydrogen atoms are different on each side. Try a 2 in front of the NaOH:

$2NaOH + H_2SO_4 \rightarrow Na_2SO_4 + H_2O$

But things are still not right. The left has two hydrogen atoms and one oxygen atom more than the right. 2 is needed in front of the H_2O:

$2NaOH + H_2SO_4 \rightarrow Na_2SO_4 + 2H_2O$

The equation is now balanced. Count the atoms and see if you agree.

Balancing a chemical equation

When an equation doesn't balance, you can fix it by putting numbers in front of the formulae, until the atoms on each side are the same. This is called **balancing the equation**. Remember that a number in front of a formula affects all the atoms in the formula. So $3H_2O$ means six hydrogens and three oxygens.

You will get plenty of practice in balancing equations when you study chemistry later on in the school. But meanwhile, see if you can do the exercises that follow.

Exercises

Write balanced equations for these reactions, using the rules you have learned. You will need to work out the formulae of the compounds, unless they are given:

1 magnesium + sulphuric acid →
$\qquad\qquad$ magnesium sulphate + water

2 hydrogen + oxygen → steam

3 sodium hydroxide + copper sulphate →
$\qquad\qquad$ copper hydroxide + sodium sulphate

4 zinc + copper chloride → zinc chloride + copper

5 hydrogen + lead oxide → lead + steam
$\qquad\qquad$ (PbO)

6 calcium carbonate + hydrochloric acid →
$\qquad\qquad$ calcium chloride + carbon dioxide + water

7 iron + chlorine → iron(III) chloride
$\qquad\qquad$ (Cl_2)

8 calcium + oxygen → calcium oxide

Professor Bunsen, naturally

Where did it all start?

The first people to use gas as a fuel for lighting and heating were probably the Chinese, over 2000 years ago. They used pigs' bladders to collect the natural gas from rotting vegetation. Then they pricked tiny holes in the bladders so that the gas could escape slowly and be burned as a flame.

In Britain, people have known for centuries that rotting vegetation produces a gas that will burn. In marshy areas, the gas was known as *Will o' the Wisp*, because of the way it escaped through the ground and burned in tiny wispy flames. Down coal mines it was called *fire damp*, and was known to explode dangerously when mixed with air.

The problem was that there was never enough of this gas for large scale use. So people tried to make it artificially. The first person to succeed was William Murdoch. In 1792 he built a furnace in which he heated coal to make gas for lighting his home. He was so successful that, 13 years later, he built a complete gasworks for a cotton mill in Manchester, and saved the mill around £2000 a year in candles – a lot of money in those days!

After that, coal gas became an accepted fuel for heating and lighting. By 1812, a company had been formed in London to supply the public with coal gas. By the middle of the 19th century the gas was being used in laboratories.

In many towns, you can still see the giant containers that were used to store coal gas. The containers floated on water and the gas entered them in just the same way that gas bubbles up through water, into a gas jar in the laboratory. Look at Figure 1.

Figure 1   A gas holder

Professor Bunsen

Robert Wilhelm Bunsen was made Professor of Chemistry at the University of Heidelberg, in Germany, in 1852. Although he discovered many things, such as an antidote for arsenic poisoning, people remember him best for his gas burning apparatus.

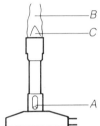

Figure 2   How the Bunsen burner works (see text)

Professor Bunsen found that, if he lit a jet of gas, it burned with a yellow, sooty flame. If he mixed air with the gas before he lit it, the mixture exploded. Somehow, he had to mix the air with the gas in a controlled way, and burn the mixture as it flowed out of a tube. So he designed his burner. Look at Figure 2. As the gas gushes out of the jet, at *A*, it drags air in through the hole. The two form a burnable mixture. At *B*, this mixture is lit. It tries to burn back down the tube, but it can't, because more gas is pushing out all the time. The flame stays on the end of the tube because the speed of burning just equals the flow of the gas. The blue cone, *C*, is unburnt gas that has not yet reached the hot, burning part of the flame.

If the air hole is left open and the gas turned right down, then the flame can burn down the tube faster than the gas can get out. When this happens the flame *strikes back* and burns at the jet. This makes the Bunsen burner very hot. It must be turned off and relit properly.

You can see that the Bunsen burner is a clever piece of apparatus. The air hole and gas jet have to be just the right size for it to operate correctly.

North Sea gas

Around 1960, geologists found that there were large quantities of natural gas under the North Sea. Over the next 15 years, millions of pounds were spent in drilling wells under the sea to detect and extract the gas. Special drilling rigs had to be designed for the rough North Sea conditions (see Figure 3), and special pipes had to be laid on the sea bed to bring the gas ashore.

Figure 3   A North Sea gas rig

Look at the table below. It shows that natural gas gives out over twice as much heat as coal gas, when it burns. In other words, it burns with a much hotter flame.

	coal gas	natural gas
heat produced per cubic metre of gas	18 million joules	39 million joules
volume of air needed to burn a cubic metre of gas	4.13 cubic metres	9.8 cubic metres

The table also shows that natural gas needs a lot more air than coal gas does, for burning. You can see why, if you compare the equations for the combustion of hydrogen and methane. A hydrogen molecule needs half an oxygen molecule (one oxygen atom) when it burns. But a methane molecule needs two oxygen molecules (four oxygen atoms) when it burns:

in coal gas $H_2 + \frac{1}{2}O_2 \rightarrow H_2O$

in natural gas $CH_4 + 2O_2 \rightarrow CO_2 + 2H_2O$

When people first tried to burn natural gas in ordinary coal-gas Bunsen burners, they found that things were not quite right. The flame was yellow and sooty because the air holes at the bottom of the tubes were not big enough. Sufficient air was not getting in. So the air holes had to be made larger to correct this.

The British Gas Corporation was keen to use the new gas, rather than coal gas, for several reasons. First of all, there was a lot of it. Second, it was a much cleaner gas that produced no harmful sulphur dioxide fumes when it burned. Third, it burned with a much hotter flame than coal gas.

Apart from the vast expense of getting the natural gas out of the sea bed and onto the shore, the Gas Corporation had another problem. A whole new network of pipelines had to be built throughout the country. Besides, gas burners of all sorts, in houses, factories, schools, and hospitals, had to be changed to suit the new gas.

Another problem concerned the gas jets. Since natural gas gives out more heat when it burns, the jets had to be made smaller to keep the heat output the same. Besides, the pressure of the gas had to be increased about three-fold, in order to draw more air in through the larger air holes.

New gas for old

Look at the information in the pie charts below. It shows how natural gas and coal gas differ.

Finally, it was found that natural gas burned more slowly than coal gas, so that the flame had an annoying habit of floating off the top of the burner. To stop this, the top was made wider and a flange added, so that the gas and air mixture came out more slowly. Look at Figure 4. This shows the differences between a natural gas burner and a coal gas burner. I wonder if Professor Bunsen would have coped with all that?

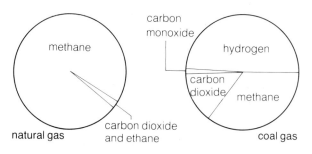

You can see that natural gas is nearly all methane, while coal gas is almost half hydrogen.

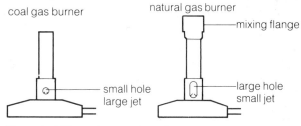

Figure 4   Different burners for different gases

Topic 3 Exercises

1. What is an atom?

2. What is a molecule?

3. Write down John Dalton's four rules about atoms.

4. About how many gold atoms would fit, side by side, on a metre ruler?

5. How many nanometres are there in a metre?

6. Explain what you understand by the Kinetic Theory.

7. Use the Kinetic Theory to explain what happens to the molecules in a piece of wax, when the wax is heated until it first melts and then boils. Use diagrams to help your explanation.

8. What is an element?

9. Describe three of the different ways that elements got their names.

10. Write down the names of these elements, using the Periodic Table at the back of the book to help you:

    Mn As Kr Cd La

11. Write down the symbols for these elements:

    vanadium potassium strontium tin plutonium

12. Look at the drawing below.

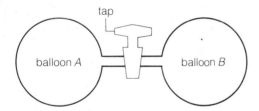

Balloon A contains hydrogen gas. Balloon B contains oxygen gas. The tap is opened and the gases mix.

a) Why do the gases mix? Why don't they stay in their separate balloons?

b) When the tap is opened, does a chemical reaction take place? If not, how could you make one happen?

c) What compound will be produced, if a chemical reaction takes place?

d) Draw diagrams to show the arrangement of the molecules in: (i) balloon A (ii) balloon B (iii) the mixture of gases (iv) the compound that would form during a chemical reaction.

13. Name the compounds with these formulae:

    $LiNO_3$ $Fe_2(SO_4)_3$ $Cu(OH)_2$ $(NH_4)_3PO_4$ $AlCl_3$

14. Write formulae for the following compounds:

calcium sulphate	aluminium sulphate
magnesium iodide	iron(III) iodide
iron(II) bromide	silver sulphate
copper nitrate	sodium carbonate
zinc oxide	ammonium nitrate

15. Write down the four things that chemical reactions have in common.

16. Put these in a sensible order:

    → products reactants

17. How would you show that:

a) a chemical reaction takes place when an iron nail is left in water?

b) a physical change takes place when salt is added to water?

18. Write balanced equations for these reactions:

a) iron + hydrochloric acid → iron(II) chloride + hydrogen

b) copper + oxygen → copper oxide

c) sodium chloride + silver nitrate → silver chloride + sodium nitrate

d) carbon + copper oxide → carbon dioxide + copper

e) potassium hydroxide + sulphuric acid → potassium sulphate + water

Things to do

19. Ask your teacher to let you try this experiment to find the melting point of wax.

a) Fill one-third of a boiling tube with candle wax.

b) Stand the tube in a beaker of boiling water until all the wax is completely melted.

c) Remove the tube from the water and fix it in a clamp.

d) Put the thermometer into the molten wax.

e) As the wax cools in the air, take its temperature every 30 seconds. *Be sure to stir the wax all the time so that it cools evenly.* Do not remove the thermometer from it.

f) Continue to take the temperature until the wax has solidified, and then for 5 minutes longer.

g) Plot a graph of wax temperature against time. Your graph should look like the one below. The melting point of the wax (or its freezing point, which is the same thing) is the temperature at which the graph flattens out.

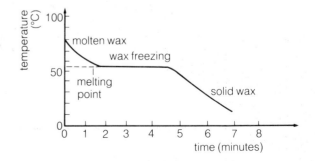

56

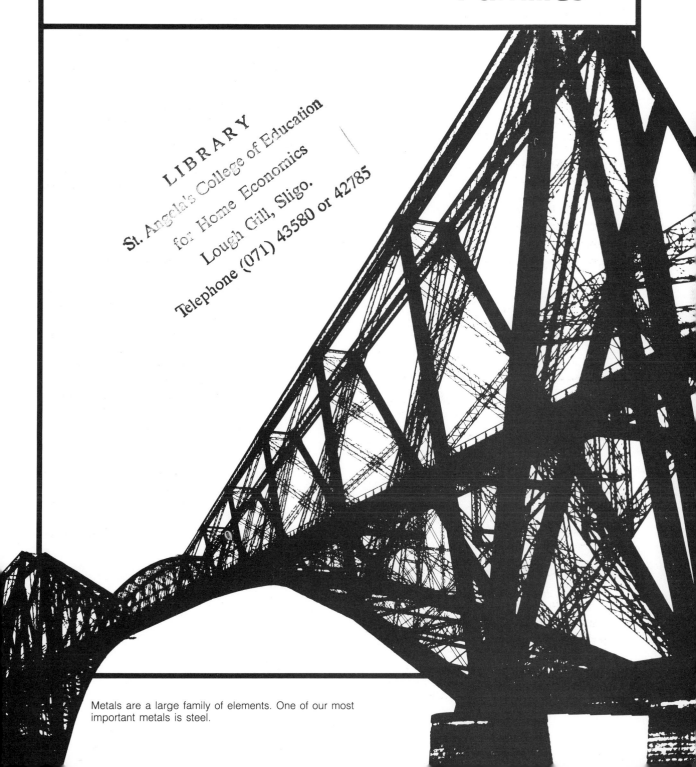

Families

Metals are a large family of elements. One of our most important metals is steel.

4.1 Acids

The acids are a family – they all usually react in the same way.
Once you know the general reactions, you can predict how an acid
will behave.

Recognising acids

The simplest acids to recognise are the common laboratory acids,
because their names all end in 'acid'. Learn their names and
formulae:

hydrochloric acid	HCl
nitric acid	HNO_3
sulphuric acid	H_2SO_4

Notice one important thing about these acids – they all contain
hydrogen. This hydrogen isn't present as a gas. It is combined with
other elements, forming compounds.

In the laboratory, acids are generally mixed with water to *dilute*
them. This is usually noted on the bottles. For example, a bottle may
have a label saying 'Dilute hydrochloric acid' or even 'Hydrochloric
acid dil.' Look at Figure 1. Concentrated acids can be dangerous
and you should never use them without your teacher's permission.

Many other substances that are used around the house, or that
occur in nature, are also acids. A few examples are shown in the
table below:

substance	chemical name
vinegar	ethanoic acid
lemon juice	citric acid
orange juice	citric acid
rhubarb juice	ethanedioic acid
ant sting	methanoic acid
stinging nettle juice	methanoic acid
vitamin C	ascorbic acid

Testing for acids

Many 'kitchen' acids can be recognised by their sour taste. Think
of vinegar, lemon juice, and rhubarb. These are safe to taste. But
remember, you must never taste anything in the laboratory.

Some natural substances change colour when they are put into
acids. Examples are blackberries, blackcurrants, and red cabbage
(which is purplish). These all go red.

Chemists use a natural substance called *litmus* to test for acids.
Litmus is extracted from a lichen that grows on trees. It can either
be dissolved in alcohol, to give a purple solution, or soaked into
paper. Look at Figure 2. Litmus is *red* in acids. If you put litmus into
a liquid and it goes red, you know the liquid is an acid.

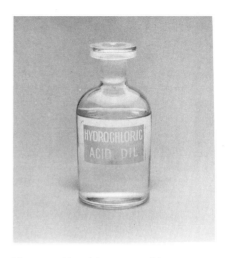

Figure 1 Most laboratory acids are
dissolved in water to dilute them.

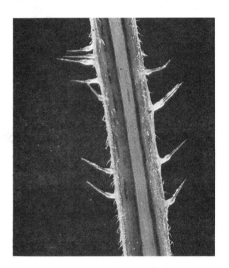

A stinging nettle barb can inject you with
methanoic acid.

Figure 2 Litmus is a vegetable extract
that can be soaked into paper. It changes
colour in acids – it goes red.

Metals and acids

Most metals react with dilute solutions of acids to produce the gas *hydrogen*. The hydrogen bubbles off. A solution of a compound called a *salt* is left behind. Here is the general word equation for the reaction:

metal + acid → hydrogen + salt

Here are two examples:

1 magnesium + hydrochloric acid → hydrogen + magnesium chloride

Mg + $2HCl$ → H_2 + $MgCl_2$

2 zinc + sulphuric acid → hydrogen + zinc sulphate

Zn + H_2SO_4 → H_2 + $ZnSO_4$

Rhubarb, vinegar, and lemons are acids which are safe to taste.

But not all metals react with acids. For example, copper, silver and gold do not react with either hydrochloric acid or sulphuric acid. Nitric acid does react with metals, but not to form hydrogen.

Carbonates and acids

Compounds ending in '*carbonate*' react with acids to produce the gas *carbon dioxide*. The gas bubbles off, and a solution of a salt is left behind. Here is the general word equation for the reaction:

carbonate + acid → carbon dioxide + salt + water

Here are two examples:

1 calcium carbonate + hydrochloric acid → carbon dioxide + calcium chloride + water

$CaCO_3$ + $2HCl$ → CO_2 + $CaCl_2$ + H_2O

2 copper carbonate + sulphuric acid → carbon dioxide + copper sulphate + water

$CuCO_3$ + H_2SO_4 → CO_2 + $CuSO_4$ + H_2O

Raw red cabbage is purple and white. When pickled in vinegar (an acid), it goes red.

Exercises

1 Write down the names and formulae of the three main laboratory acids.

2 What do acids do to litmus?

3 What would you expect to see when a piece of calcium is put into dilute hydrochloric acid? Name the two substances that form.

4 What would you expect to see when zinc carbonate is added to dilute nitric acid? Name the substances that form.

Things to do

Collect as many different berries and coloured vegetables as you can. Examples are blackberries, sloes, raspberries, beetroot, red cabbage, carrots, and so on. Squash each sample onto a piece of filter paper so that its colour stains the paper. Then test the coloured paper with an acid like vinegar. Does the paper change colour in acid? Which berries and vegetables could be used to test for acid?

4.2   A closer look at hydrogen

Hydrogen is the lightest of all gases, but this can be a problem.

Making hydrogen

On the last page you saw that when metals react with dilute
acids, hydrogen gas is formed. Figure 1 shows the apparatus that
would be used to make and collect a sample of hydrogen in the
laboratory. Dilute hydrochloric acid is poured onto pieces of zinc in
a flask. Hydrogen bubbles out of the flask, flows along the delivery
tube, and bubbles up through water into a gas jar. The water in the
gas jar is pushed down by the hydrogen, until the jar is full of gas.
(This apparatus can also be used to make and collect other gases,
as you will see on page 62.)

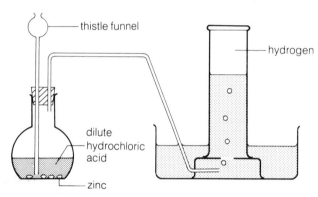

Figure 1   The laboratory preparation of hydrogen

The properties of hydrogen

Hydrogen is an invisible gas, with no taste or smell. What makes it
very interesting is that it is the lightest of all gases. Its molecules
are very tiny and can move faster than any other gas molecules.
So when a gas jar full of hydrogen is opened to the air, the gas
escapes very quickly and mixes through the air. As you saw in Unit
3.4, this mixing is called **diffusion**.

Look at the apparatus in Figure 2. The porous pot is made of
unglazed pottery. Air molecules continually pass in and out of
the tiny holes in it. If a beaker is put over the pot and filled with
hydrogen, then hydrogen molecules pass in through the holes in the
pot while air molecules pass out. Because the hydrogen molecules
are so small and fast moving, they get into the pot faster than the
air molecules can get out. The result is that pressure builds up
inside the pot, and bubbles are forced out of the end of the tube.

Burning and exploding

If a jet of hydrogen gas is lit, it burns with a tiny, almost invisible
flame. Look at the equation for the reaction:

hydrogen + oxygen $\rightarrow$ steam
$$2H_2 \quad + \quad O_2 \quad \rightarrow 2H_2O$$

The reaction gives out a lot of energy. Note that the product is

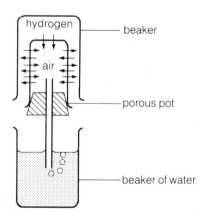

Figure 2   Hydrogen molecules diffuse
into the porous pot faster than air
molecules diffuse out. Pressure builds up
inside the pot and bubbles are forced out
of the end of the tube.

steam. When the steam condenses, pure water is formed. See Figure 3. Some scientists think that in the next century, when petrol will be scarce and very expensive, hydrogen will be the ideal fuel for cars and lorries. There will be no more exhaust fumes, just little puffs of steam!

However, hydrogen can be dangerous. If it is mixed with air or oxygen before being lit, the mixture will explode. The reaction is the same as before, but it happens much more quickly. The steam forms so rapidly that it pushes back the air, causing a shock wave. We hear this as a bang.

A test for hydrogen

This test depends on the facts that hydrogen mixes quickly with air and that the mixture explodes when lit. Look at Figure 4. As soon as the bung is taken out of a test tube of hydrogen, some of the gas diffuses out and some air diffuses in. In less than a second, an explosive mixture is made. When a lighted splint is held to the mouth of the test tube, the mixture explodes with a small squeaky 'pop'.

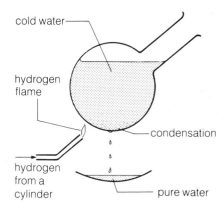

Figure 3 When hydrogen burns, steam is produced. The steam can be condensed to give pure water.

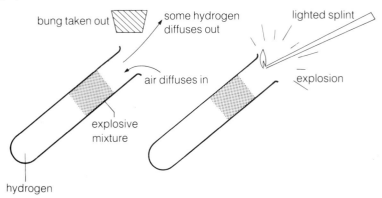

Figure 4 If the lighted splint goes 'pop', then the gas in the test tube is hydrogen.

Uses for hydrogen

Hydrogen is used in the manufacture of many everyday things. It is reacted with nitrogen to make *ammonia*. The ammonia is then used to make *fertilizers*. Many *plastics* such as PVC need hydrogen in the early stages of their production. Hydrogen is also used to turn liquid vegetable oils into *margarine*. See Figure 5.

Figure 5 Hydrogen is used in the manufacture of margarine. It helps to solidify the vegetable oils.

Exercises

1 Zinc and dilute hydrochloric acid can be used to make hydrogen. Name another metal and another acid that could be used instead. Write a balanced equation for the reaction that takes place.

2 Fifty years ago, people were flown across the Atlantic in airships filled with hydrogen. What property of hydrogen enables it to lift heavy loads? What property made these airships dangerous?

The manufacturers stopped using hydrogen after several disastrous accidents. In modern balloons, another gas is used instead. Find out what this gas is. Why it is much safer than hydrogen?

3 Describe how you would test a sample of gas to see if it was hydrogen.

4 Explain what is meant by the term 'diffusion'.

4.3　A closer look at carbon dioxide

Unlike hydrogen, carbon dioxide is a very heavy gas.

Making carbon dioxide

You saw on page 60 that carbon dioxide is made whenever a carbonate reacts with an acid. In the laboratory, *calcium carbonate*, in the form of marble, is used for making the gas. Look at Figure 1. The apparatus is the same as that for making hydrogen. Carbon dioxide is produced when the hydrochloric acid runs onto the chips of marble, and it is collected by bubbling it through water into a gas jar.

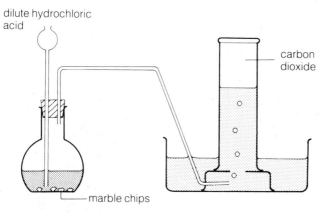

Figure 1　The laboratory preparation of carbon dioxide

This is the equation for the reaction:

$$\begin{array}{ccccccccc} \text{calcium} & + & \text{hydrochloric} & \rightarrow & \text{carbon} & + & \text{calcium} & + & \text{water} \\ \text{carbonate} & & \text{acid} & & \text{dioxide} & & \text{chloride} & & \\ CaCO_3 & + & 2HCl & \rightarrow & CO_2 & + & CaCl_2 & + & H_2O \end{array}$$

The properties of carbon dioxide

Like hydrogen, carbon dioxide has no colour or taste or smell. But unlike hydrogen, it is much heavier than air. Figure 2 shows the 'invisible gas trick'. Gas jar *A* contains air, so the candle burns. Gas jar *B* contains carbon dioxide. The carbon dioxide from *B* can be poured into *A* because it is heavier than air. The candle goes out at once. Flames cannot continue to burn in carbon dioxide.

Testing for carbon dioxide

1　Carbon dioxide *does not support combustion*. That means that things will not burn in it. So, if a burning splint is put into a test tube of carbon dioxide, the flame goes out at once.

2　When carbon dioxide is bubbled through a solution of calcium hydroxide, a suspension of tiny particles of insoluble calcium carbonate is formed. This makes the solution go cloudy. The common name for calcium hydroxide solution is *lime water*. We say that carbon dioxide makes lime water go cloudy. See Figure 3.

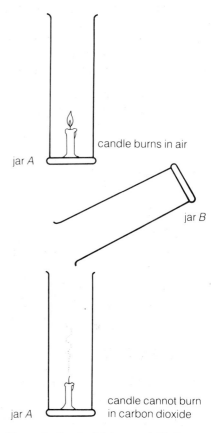

Figure 2　The 'invisible gas trick'. The candle will burn in gas jar *A* because it contains air. It will not burn when carbon dioxide is poured onto it from gas jar *B*.

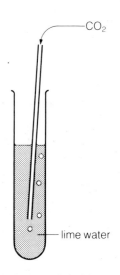

Figure 3　Carbon dioxide turns lime water cloudy.

Uses for carbon dioxide

Carbon dioxide is only slightly soluble in water at ordinary pressure, but it dissolves very well when the pressure is increased a little. Fizzy drinks contain dissolved carbon dioxide. While the top is on the bottle, the gas remains dissolved under pressure. When the top is removed, the pressure is reduced. The carbon dioxide comes bubbling out, giving the drink a sparkle. See Figure 4.

When carbon dioxide gas is cooled to below −78 °C, it freezes to a white, ice-like solid. This is called *dry ice*. Dry ice is good for keeping frozen food frozen, because it is much colder than ordinary ice. See Figure 5.

Figure 4   Carbon dioxide puts the sparkle into fizzy soft drinks.

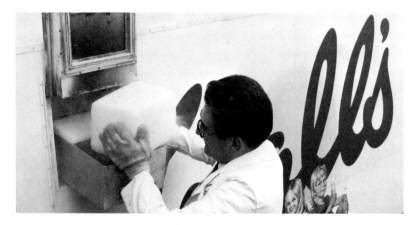

Figure 5   Solid carbon dioxide is called dry ice. It can be used to keep things cold.

You saw earlier that carbon dioxide is heavier than air, and that it does not support combustion. This means it can be used to put out fires. It does so by forming a blanket over the fire. This cuts off the air supply and the fire goes out. Carbon dioxide is particularly useful for putting out electrical fires or chemical fires, where water could be dangerous. Figure 6 shows a carbon dioxide fire extinguisher. The carbon dioxide is under pressure in the cylinder. When the lever is pressed, the pressure is released and the gas comes out with force.

Figure 6   Putting water on an electrical or chemical fire can be dangerous. A carbon dioxide fire extinguisher should be used. Is there one in your laboratory?

Exercises

1   Calcium carbonate and hydrochloric acid react together to form carbon dioxide. Name another carbonate and another acid that could be used instead. Write a balanced equation for the reaction that takes place.

2   Look at the apparatus shown on the right. What do you think will happen? It may help you to look back at the properties of hydrogen, on page 60. And remember, carbon dioxide is heavier than air.

3   Why does a bottle of fizzy drink go 'flat' if it is left open?

4   Describe two tests that you would perform on a gas to show that it was carbon dioxide and not hydrogen.

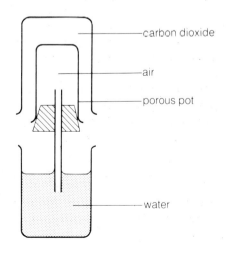

4.4 Bases, indicators, and pH

Bases are the chemical opposites of acids. But how do you tell acids and bases apart? Read on....

What is a base?

Any compound that starts with the name of a *metal*, and ends in *'oxide'* or *'hydroxide'*, is a *base*. Bases are oxides and hydroxides of metals. Look at the compounds in Figure 1. They are all bases.

Some bases dissolve in water. These have a special name. They are called *alkalis*. Alkalis are bases that dissolve in water. The alkalis that are generally used in the laboratory are shown in Figure 2.

Recognising alkalis

1 Some alkalis, like sodium hydroxide and potassium hydroxide, feel *soapy* when you get them on your fingers. This is because they react with the fats in your finger tips to make compounds similar to soap. Real soap is made by boiling potassium hydroxide with oils or fats, such as palm oil or beef fat. This process is called *saponification*. Look at Figure 3.

2 Like acids, alkalis change the colour of certain plant compounds. Chemists use litmus to test for alkalis as well as for acids. Litmus is *blue* in alkalis. (Remember, it is red in acids.)

Indicators

Litmus is called an *indicator*. It tells you whether the substance you are testing is an acid or an alkali, by the colour it goes. Many fruits and vegetables like blackberries, blackcurrants, red cabbage, and beetroot also change colour in acids and alkalis, but their colours fade. Litmus is more reliable. But litmus is not the only indicator used in the laboratory. Compare these:

indicator	colour in acid	colour in alkali
litmus	red	blue
methyl orange	red	yellow
phenolphthalein	colourless	red

Strengths of acids and alkalis

Acids and alkalis are of different *strengths*. This isn't anything to do with how concentrated or dilute they are. It is about how good they are at being an acid or an alkali.

For example, citric acid and ethanoic acid are both *weak acids*. They are weak enough for you to be able to drink them in soft drinks and vinegar. (But remember, you should *never* drink them in the laboratory.) On the other hand, hydrochloric acid, sulphuric acid, and nitric acid are *strong acids*. They are far too corrosive to drink even when diluted with water. Look at Figure 4.

copper oxide	CuO
zinc oxide	ZnO
magnesium oxide	MgO
iron (III) oxide	Fe_2O_3
aluminium hydroxide	$Al(OH)_3$
lead hydroxide	$Pb(OH)_2$

Figure 1 Some examples of bases

sodium hydroxide	NaOH
potassium hydroxide	KOH
calcium hydroxide (lime water)	$Ca(OH)_2$
ammonium hydroxide	NH_4OH

the ammonimum group, NH_4, acts like a metal

Figure 2 These are the alkalis generally used in the laboratory.

Figure 3 Oil from palm trees is boiled with potassium hydroxide to make soap.

not safe to drink   safe to drink

Figure 4 Weak acids are safe to drink. Strong acids diluted with water are still strong and are not safe to drink.

The same is true of alkalis. Potassium hydroxide and sodium hydroxide are strong alkalis. They will dissolve the tips of your fingers and are dangerous, *caustic* substances. Ammonium hydroxide and calcium hydroxide are much weaker, but you still could not drink them safely.

However, magnesium hydroxide is a very weak alkali, and is used in Milk of Magnesia and other indigestion mixtures. It is not soluble enough to be used as a laboratory alkali, but it will turn litmus blue. Look at Figure 5.

pH numbers

The strength of an acid or alkali is shown by its *pH number*. The term 'pH' comes from German words meaning 'strength of acid'. Look at the scale below:

Figure 5 Only weak alkalis are safe to drink. Indigestion mixtures contain a very weak alkali (magnesium hydroxide).

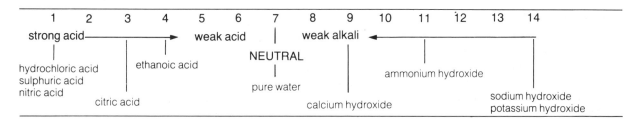

Measuring pH numbers

To measure the pH number of an acid or an alkali, chemists use:

1 *A pH meter.* This is an electrical instrument that measures the pH through an electrode and shows it on a scale. Look at Figure 6.

2 *Universal indicator.* This is a mixture of several indicators. The indicators are chosen so that the mixture goes a different colour at different pH numbers. Universal indicator can be used as a solution, or soaked onto paper. There are many different types, and the colours vary slightly from type to type. A colour chart is supplied with universal indicator, showing the colour changes. The chemist puts the indicator into the acid or alkali and compares the colour of the solution with the colour chart, to see what the pH number is. Here is an example of a colour chart:

Figure 6 A pH meter is an electrical instrument which measures pH.

pH number	1	2	3	4	5	6	7	8	9	10	11	12	13	14
colour	red	pink		beige	yellow		green		dark green			light blue		dark blue

Exercises

1 What is a base? Write down the names of four bases not mentioned on these pages.

2 What do chemists call a base that dissolves in water?

3 Why do chemicals like sodium hydroxide and potassium hydroxide feel soapy to touch?

4 How is soap made?

5 What is an indicator? Name an indicator other than litmus and explain how you would use it.

6 Explain what the pH number tells you about an acid or an alkali.

7 What is universal indicator?

8 What is Milk of Magnesia used for? Why is it safe to drink, while sodium hydroxide is very dangerous?

9 Find out the pH numbers of these liquids:
   a) lemonade b) rain water c) oven cleaner.

4.5 Salts

A salt is made whenever an acid is neutralised by a base. Almost every metal can be made into a variety of salts.

Everyday salts

It is confusing that the white, crystalline solid, that makes crisps and chips taste so much better, is known as salt. In fact it is only one member of a huge family of compounds called *salts*. Figure 1 shows some everyday salts.

Naming salts

Salts are made from acids, and their names come from the names of the acids.

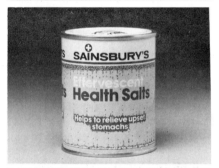

name of acid	name of salt	example
hydrochloric acid, HCl	chloride	sodium chloride, NaCl calcium chloride, $CaCl_2$
nitric acid, HNO_3	nitrate	potassium nitrate, KNO_3 ammonium nitrate, NH_4NO_3
sulphuric acid, H_2SO_4	sulphate	copper sulphate, $CuSO_4$ magnesium sulphate, $MgSO_4$
carbonic acid, H_2CO_3	carbonate	calcium carbonate, $CaCO_3$
phosphoric acid, H_3PO_4	phosphate	sodium phosphate, Na_2HPO_4

Neutralisation

When a base reacts with an acid, a salt is formed. Water is always formed as well. This process is called *neutralisation*. Look at an example:

$$\text{copper oxide} + \text{sulphuric acid} \rightarrow \text{copper sulphate} + \text{water}$$
$$CuO + H_2SO_4 \rightarrow CuSO_4 + H_2O$$

Notice that the salt is called *copper sulphate* because it is made from the base *copper* oxide and the acid *sulph*uric acid.

Look at the equation again. The oxide part of the base has reacted with the hydrogen part of the acid to form water. Now check the following example to see if the same thing has happened:

$$\text{potassium hydroxide} + \text{hydrochloric acid} \rightarrow \text{potassium chloride} + \text{water}$$
$$KOH + HCl \rightarrow KCl + H_2O$$

This time the hydroxide part of the alkali has reacted with the hydrogen part of the acid to form water. *Whenever neutralisation takes place, a salt and water are formed:*

$$\text{acid} + \text{base (or alkali)} \rightarrow \text{salt} + \text{water}$$

Figure 1 Some everyday salts

66

Salts from other substances

You have learned already that metals and carbonates also react with acids to form salts. Look at these examples:

iron	+	sulphuric acid	→	iron(II) sulphate	+	hydrogen
Fe	+	H_2SO_4	→	$FeSO_4$	+	H_2
metal	+	acid	→	salt	+	hydrogen

calcium carbonate	+	nitric acid	→	calcium nitrate	+	carbon dioxide	+ water
$CaCO_3$	+	$2HNO_3$	→	$Ca(NO_3)_2$	+	CO_2 carbon	+ H_2O
carbonate	+	acid	→	salt	+	dioxide	+ water

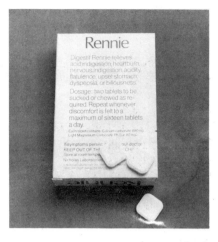

Figure 2   These salts neutralise excess acidity in the stomach.

Neutralisations all around you

On many farms, the soil is too acidic to let crops grow properly. So lime (calcium oxide) is spread on the fields to neutralise the excess acidity.

In your stomach, indigestion tablets that contain sodium hydrogencarbonate, calcium carbonate, or magnesium hydroxide, will neutralise the excess acid that causes you pain. Figure 2 shows the salts in one brand of indigestion tablets.

Health salts often contain a mixture of sodium hydrogencarbonate and tartaric acid. When water is added, the acid reacts with the carbonate and a fizz of carbon dioxide is produced.

Sometimes people dab a bee sting with sodium hydrogencarbonate solution, because the sting is acidic. Similarly, they treat a wasp sting, which is alkaline, with a weak acid like vinegar or lemon juice.

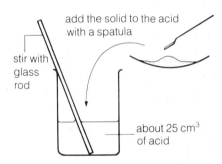

Figure 3   To make a salt, neutralise an acid by adding a base.

A method for making a salt

Here is how to make a salt from an acid, starting with an insoluble base, a metal, or a carbonate.

1   If a metal or carbonate is used, the acid can be cold. If an insoluble base is used, the acid must first be heated over a Bunsen burner. Add the metal, carbonate, or base to the acid until no more will dissolve. That means that all the acid has been neutralised. Look at Figure 3.

2   Filter the liquid, as shown in Figure 4, to remove the unreacted metal, carbonate, or base. The residue can be thrown away. The filtrate is a solution of the salt.

3   Evaporate the filtrate so that some of the water is boiled away and small crystals appear around the edge of the basin. Then leave the solution in a cool, clean place so that it can crystallise.

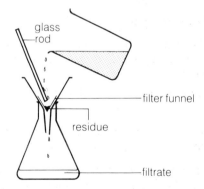

Figure 4   Filter the mixture of acid and base. The filtrate is a solution of the salt.

Exercises

1   Name three different types of substances that react with acids to form salts.

2   What is meant by the term 'neutralisation'? What two substances are always formed when neutralisation takes place?

3   Name the salt that will be formed from:
   a) aluminium and hydrochloric acid
   b) zinc carbonate and nitric acid
   c) iron(III) oxide and sulphuric acid.

4.6 The Periodic Table

In science, you will meet many lists and diagrams. The Periodic Table is one of the most important because it contains all the elements that exist.

A list of elements

You will remember that an element is a pure substance that contains only one sort of atom. There are 92 naturally-occurring elements and a further 15 or so that are man-made. The Periodic Table is a list of these elements. But it is not just a straight list. It is arranged in a special way, as shown on page 70.

The position of an element in the Periodic Table tells chemists a great deal about the structure of its atoms, what the element is like, and how it will react. In other words it tells them about the *properties* of the element.

The vertical columns in the Periodic Table are called *groups*. They are numbered from I to VII, and then the last one is called Group 0. The elements in a group all have very similar properties, as you will see below.

Group I   The alkali metals

Figure 1 shows you the elements in Group I, and their symbols. The Group I elements are often called the *alkali metals*. Your teacher can probably show you samples of lithium, sodium, and potassium. Rubidium and caesium are not kept in the laboratory, because they are expensive and difficult to store, and francium is radioactive and has only been found in tiny amounts.

The Group I elements all have very similar properties:
1   Lithium, sodium and potassium are all light enough to float on water.
2   They are all soft enough to be cut with a knife.
3   They all corrode quickly in air, so have to be kept in oil.
4   Although they are dull and corroded on the outside, they are all shiny on the inside. Look at Figure 2.
5   They all react violently with water, producing hydrogen gas and an *alkaline* solution. This is why they are called 'alkali metals'. Potassium reacts so violently that the hydrogen catches fire. Rubidium and caesium react almost explosively. All the elements except lithium melt when they react with water.
6   When they are heated, the alkali metals melt at a low temperature and burn with a coloured flame. They react with oxygen in the air to form compounds called *oxides*. Lithium burns with a crimson flame. The sodium flame is golden yellow and the potassium flame is lilac.

Group VII   The halogens

Figure 3 shows you the elements in Group VII, and their symbols. These elements are often called the *halogens*. They all react very

lithium	Li
sodium	Na
potassium	K
rubidium	Rb
caesium	Cs
francium	Fr

Figure 1   The elements in Group I are called the alkali metals.

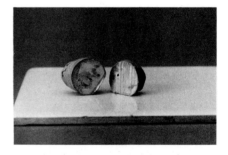

Figure 2   Group I elements are shiny on the inside. This photo shows a freshly cut piece of sodium.

fluorine	F
chlorine	Cl
bromine	Br
iodine	I
astatine	At

Figure 3   The elements in Group VII are called the halogens.

easily with metals, forming *salts*. 'Halogen' is a Greek word meaning 'salt producer', so you can see how they got their name.

Like the alkali metals, the halogens have many properties in common:

1 They are all coloured. Fluorine is a yellow gas, and chlorine a pale green one. Bromine is a blood-red liquid that gives off a reddish vapour. Iodine exists as shiny, dark grey crystals. When these are heated, they give off a purple vapour.

2 They are all very poisonous and very corrosive. Fluorine is too dangerous to use in the laboratory. Chlorine was used as a poison gas in World War I. It is always kept in a container marked with a warning. Bromine can cause bad skin burns, and its vapour is choking and dangerous. Iodine vapour is poisonous.

3 When dissolved in water, the halogens all act as bleaches and will take the colour out of some materials. Their solutions can also be used as antiseptics if they are dilute enough. Look at Figure 4. At swimming baths and water works, chlorine is added to the water to kill harmful bugs. Iodine solution used to be put on cuts and grazes, to sterilize them.

Figure 4 Solutions like TCP contain compounds with chlorine in them.

The transition elements

The block of elements in the middle of the Periodic Table is called the *transition elements* and it contains the *transition metals*. You can read more about the properties of metals in Unit 3.5, but look now at some of the things these metals have in common:

1 They are all fairly heavy.

2 They are all reasonably strong. Think of iron.

3 Many of them, like gold, silver, chromium and platinum, are resistant to corrosion. Look at Figure 5.

4 Many of them form coloured compounds. For example, copper sulphate is blue, iron(II) sulphate is green, and nickel chloride is dark green. Compounds that do not contain transition elements are nearly always colourless.

Figure 5 Gold, silver, chromium, and platinum are transition elements which are resistant to corrosion. They are commonly used to make jewellery.

Exercises

1 Using the Periodic Table, write down the names and symbols of the elements in Groups II and VI.

2 Describe how you would expect a small piece of potassium to behave, when it is placed in a beaker of water.

3 Why shouldn't sodium be kept open to the air?

4 Name the compound that lithium forms when it is heated in air.

5 Give one use for chlorine.

6 What may happen to a piece of coloured material, when it is soaked in an aqueous solution of chlorine?

7 What colour is iodine vapour?

8 What properties of iron make it suitable for things like girders and motor cars?

9 Can you think of an everyday use of chromium?

10 See if you can find out the name of the transition metal that is used to make blue coloured glass.

Group

non-metals

metals

	Group	
I	II	

Group

III IV V VI VII O

transition elements

1 H hydrogen																		4 2 He helium
7 3 Li lithium	9 4 Be beryllium											11 5 B boron	12 6 C carbon	14 7 N nitrogen	16 8 O oxygen	19 9 F fluorine	20 10 Ne neon
23 11 Na sodium	24 12 Mg magnesium											27 13 Al aluminum	28 14 Si silicon	31 15 P phosphorus	32 16 S sulphur	35·5 17 Cl chlorine	40 18 Ar argon
39 19 K potassium	40 20 Ca calcium	45 21 Sc scandium	48 22 Ti titanium	51 23 V vanadium	52 24 Cr chromium	55 25 Mn manganese	56 26 Fe iron	59 27 Co cobalt	59 28 Ni nickel	64 29 Cu copper	65 30 Zn zinc	70 31 Ga gallium	73 32 Ge germanium	75 33 As arsenic	79 34 Se selenium	80 35 Br bromine	84 36 Kr krypton
85 37 Rb rubidium	88 38 Sr strontium	89 39 Y yttrium	91 40 Zr zirconium	93 41 Nb niobium	96 42 Mo molybdenum	98 43 Tc technetium	101 44 Ru ruthenium	103 45 Rh rhodium	106 46 Pd palladium	108 47 Ag silver	112 48 Cd cadmium	115 49 In indium	119 50 Sn tin	122 51 Sb antimony	128 52 Te tellurium	127 53 I iodine	131 54 Xe xenon
133 55 Cs caesium	137 56 Ba barium	139 57 La lanthanum	178·5 72 Hf hafnium	181 73 Ta tantalum	184 74 W tungsten	186 75 Re rhenium	190 76 Os osmium	192 77 Ir iridium	195 78 Pt platinum	197 79 Au gold	201 80 Hg mercury	204 81 Tl thallium	207 82 Pb lead	209 83 Bi bismuth	210 84 Po polonium	210 85 At astatine	222 86 Rn radon
223 87 Fr francium	226 88 Ra radium	227 89 Ac actinium															

140 58 Ce cerium	141 59 Pr praesodmium	144 60 Nd neodimium	147 61 Pm promethium	150 62 Sm samarium	152 63 Eu europium	157 64 Gd gadolinium	159 65 Tb terbium	162 66 Dy dysprosium	165 67 Ho holmium	167 68 Er erbium	169 69 Tm thulium	173 70 Yb ytterbium	175 71 Lu lutecium
232 90 Th thorium	231 91 Pa protoactinium	238 92 U uranium	237 93 Np neptunium	242 94 Pu plutonium	243 95 Am americium	247 96 Cm curium	247 97 Bk berkelium	251 98 Cf californium	254 99 Es einsteinium	253 100 Fm fermium	256 101 Md mendelevium	254 102 No nobelium	257 103 Lw lawrencium

The periodic table of elements

4.7 Metals

You probably know what metals look like, but as a scientist you also need to know what sorts of reactions they take part in.

Comparing metals and non-metals

How can you tell whether an element is a metal?

The simple answer to this is to look at the Periodic Table on page 70. The metals lie to the left of the zig-zag line (below it) and the non-metals lie to the right of it (above it).

Figure 1 Metals can be bent or cut easily. This means they can be fashioned into decorative items.

You can also tell by studying the properties of the element. Metals and non-metals have quite different properties. Look at the list below:

metals	non-metals
They are all solids (except mercury).	They can be solids or gases (except bromine).
They can all be bent or cut easily. Look at Figure 1.	They are brittle and snap or shatter.
They are generally shiny inside and corroded on the outside.	They are not shiny or corroded.
They are *malleable*. This means they can be hammered flat. Look at Figure 2.	Non-metals are too brittle to be hammered or stretched.
They are *ductile*. This means they can be stretched. Look at Figure 3.	
They are *sonorous*. This means they clang if you hit them. Look at Figure 4.	Non-metals are not sonorous.
They allow heat and electricity to pass through them easily – they are good **conductors** of heat and electricity.	They are bad conductors of heat and electricity (except for graphite). They are called *insulators*.

Figure 2 Gold can be hammered into very thin sheets. These men are making gold leaf ten-millionths of a centimetre thick.

Figure 4 Metals clang when they are hit together.

Figure 3 Wire is made by stretching metals. This photo shows coils of newly made copper wire.

Differences in structure

In all metals except mercury, the atoms are held tightly together in regular rows and columns. Look at Figure 5. This regular arrangement is called a *metallic lattice*. It extends in all directions, resulting in a metal crystal. It may surprise you that metals form crystals, but look at Figure 6. It shows a tiny part of the surface of a piece of copper.

When a metal is bent or hammered or stretched, the rows of metal atoms are able to move over each other, without the structure coming apart. Similarly, heat and electrical energy can easily be passed from one atom to another because the atoms are close together. They are good conductors of heat and electricity.

On the other hand, non-metals do not have metallic lattices. Solids like phosphorus, iodine, and sulphur are made of molecules. Look at Figure 7. Their molecules are held together in a crystal lattice, but not as strongly as metal atoms are held in a metallic lattice. When they are bent or hammered or stretched, the crystal lattice is easily broken.

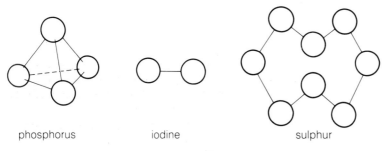

phosphorus   iodine   sulphur

Figure 7   The molecules of non-metals do not form a regular metallic lattice.

Non-metals are not good conductors of heat or electricity, because heat energy and electrical energy are not easily passed from molecule to molecule.

The odd-man-out is graphite. Graphite is one form of the non-metal carbon. It has a special structure that allows it to be a good conductor of electricity.

Reactions of metals with air

When metals are left open to the air, most of them *corrode*. This means that they combine with the oxygen in the air to form a surface coating of oxide. For example:

aluminium   +   oxygen   →   aluminium oxide
$$4Al \quad + \quad 3O_2 \quad \rightarrow \quad 2Al_2O_3$$

The coating of aluminium oxide is useful because it protects the aluminium from any further attack. Similarly, copper on a roof gains a protective coating of green *verdigris* as the copper reacts with the oxygen and carbon dioxide in the air. Iron corrodes or *rusts* in damp air. Look at Figure 8. But this corrosion does not protect the iron in any way. In fact, rust only encourages more rust.

Some metals, like sodium and potassium, corrode in air so quickly that they have to be kept under oil to protect them. A few metals,

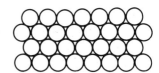

Figure 5   The atoms in a metal are arranged in a very regular way. The arrangement shown here is called hexagonal close packing.

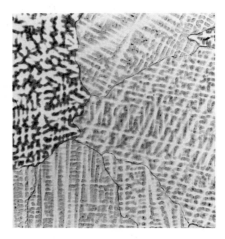

Figure 6   Metals form crystals. This magnified photo shows the crystals in a piece of copper.

Figure 8   Iron corrodes or rusts in damp air.

like gold and platinum, do not corrode in air at all. This makes them useful for jewellery, because they stay shiny, and for important electrical connections which must not get corroded away.

Metals that react with air to form oxides will react much more readily in pure oxygen, and even quicker when they are heated.

Because different metals react with oxygen at different speeds, or with different *reactivities*, they can be arranged in a league table called the *Activity Table*. Look at Figure 9.

Reactions of metals with water

The Activity Table can also be used to show how well metals react with water. Look at Figure 9 again. Metals like sodium and potassium react violently with water to form hydrogen and a solution of an alkali. This was described on page 68. Magnesium reacts only very slowly with water, but very readily with steam. Lead only just reacts with steam, when the lead is white-hot. The metals copper, silver, and gold have no reaction with water in any form. Your water pipes wouldn't last long if you made them of magnesium, but copper lasts because it does not react. You could use gold instead if you could afford it!

metal	reactivity with oxygen	reactivity with water
K	fast	react with water
Na		
Ca		
Mg		
Al		react with steam
Zn		
Fe		
Pb		
Cu	slow	no reaction
Ag	not at all	
Au	not at all	

decreasing reactivity (arrow pointing down)

Figure 9 The Activity Table is a way of arranging metals to show their different reactivities.

Exercises

1 Make a rough sketch of the Periodic Table and show where the metals and non-metals are to be found.

2 Describe three ways in which metals differ from non-metals.

3 What do the words *malleable*, *ductile*, and *sonorous* mean?

4 What does the term 'good conductor of electricity' mean? Which non-metal is a good conductor of electricity?

5 Why is sodium stored under oil?

6 What compound is formed when calcium is heated in air? Write an equation for the reaction. Write a word equation first, and then work out the formulae and write a balanced equation.

7 Gold is said to be a noble metal. Can you guess why it's called that? (Look at the way it reacts with air and water.)

8 Look at the position of aluminium in the Activity Series. Will it react with cold water?

9 Why are water pipes made of copper rather than iron?

Bells

Not just for fire alarms

One property that distinguishes metals from other materials is that they clang when you strike them. Metals are said to be sonorous. You probably enjoy the sound you make when you bang on a metal dustbin lid, and if you drag a stick along a line of iron railings the effect can be most tuneful. By contrast however, the sound of two cars crashing together is not very musical at all. The difference lies in the shape of the metals and the way in which they are hit.

The clang made when you strike a metal is caused by the metal vibrating. The *vibrations* produce energy waves in the air. When these strike your ear drums you 'hear' a sound. Different shapes and thicknesses of metal will vibrate in different ways, and so produce different sounds. Usually a mixture of metals is used in musical instruments, and in bells. The sounds from these will also depend on which metals are used in the mixture.

Think of a fire alarm bell and a church bell. Both ring and both draw attention to themselves, but the church bell has a much more pleasing sound.

Most musical instruments make mixtures of notes consisting of a pure *fundamental* note, and *harmonics*, which are other notes at varying heights above the fundamental. The depth and richness of the sound from a bell depends on these harmonics.

After a bell has been cast, the bell founder *tunes* it by shaving off thin layers of metal from different places inside it, using an electrically-operated lathe. Before this machine was invented, steam power was used to cut the metal. Before that it was removed by chipping away at the rim. When the bell is properly tuned, its voice consists of five notes: the *strike note*, which is heard when the bell is first struck; the *hum note*, which is an octave lower and gradually swells up; and three other harmonics. Much better than a fire alarm!

Some history of bells

Bells have been used for worship in churches and temples, and as musical instruments, for hundreds if not thousands of years. It was probably the Romans who brought them to this island. As early as the 7th century, large church bells were being made, and by the 13th century there were about 300 bells in English church towers. Very often in towns and villages you can spot where bells were once made, from the names of streets, such as 'Bell Lane', or of fields, such as 'Bell Meadow'. Often a big bell, too large to transport, would be made right on the spot, in a meadow next to the church.

In Gloucester, an ancient Roman city, there are references to bell founding from as early as 1270. And starting in the 1680s, the Rudhall family of Gloucester made bells for over 150 years, handing the skill on through four generations. In this period, some 5000 bells were cast.

In London, at the Whitechapel Bell Foundry, bells have been made for over 400 years. They hang in many famous churches, as well as unknown ones, both in this country and abroad. Big Ben and the Liberty Bell in the USA are both Whitechapel bells. Look at Figure 1.

Figure 1 The Great Bell of Bow is a Whitechapel bell. It hangs in St Mary le Bow Church in London.

Casting bells

The shape of a bell is most important because it determines the way the bell will sound. Look at Figure 2. You can see that the inside and outside shapes are different. The bell is thickest at the *soundbow*, where it is struck by the clapper. To achieve the correct shape, separate moulds are made for the inside and outside of the bell.

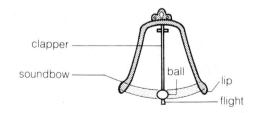

Figure 2 The shape of a bell affects its sound.

First of all, the *core* is made by packing clay over bricks. The clay is scraped into shape by a *gauge*, and the final shape is coated in graphite after it has been dried in an oven. Look at Figure 3.

Figure 3 The core is moulded by packing clay over bricks.

Next, an outer iron mould called a *cope* is lined with clay and shaped and dried in the same way. If the bell is to have an inscription, the letters must be stamped on the inside of the cope, back to front, so that they will appear on the outside of the bell the right way round.

The next stage is the crucial one. The cope is carefully lowered over the core and the two are clamped together. The bell metal is prepared by melting together copper and tin to make an alloy. (The alloy contains 77% copper.) The amount has to be estimated carefully. Sufficient alloy must be made to fill the mould in one go, and it would be wasteful to make too much. A bell with a diameter of 100 cm will contain about 550 kg of metal, and this will cost at least £1000.

When the molten metal is ready, it is poured into the space between the cope and core (Figure 4).

Figure 4 The bell is cast by pouring molten metal into the space between the cope and the core.

This process is called *casting* the bell. At this stage, if the clay has not been dried properly, steam will form explosively and the mould will shatter. At last, when the metal is cool, the two halves of the mould are removed and the bell is cleaned and tuned.

Ringing bells

When a church bell is rung correctly, it is swung in a full circle so that it strikes the clapper as it swings. To get the bell ready for ringing, the ringer swings it from side to side, pulling on the rope so that it swings higher every time. When it swings high enough, it can balance upside-down.

Bells ringing in Guildford Cathedral

When the bell is to be rung, the ringer pulls on the rope. The bell topples over, hitting the clapper as it swings. When it has reached the far extent of its swing, the ringer pulls on the rope to make it swing back again. He can make it swing quickly or slowly, depending upon how high he lets it swing each time. This is important, for the ringer must be able to make the bell ring at different intervals to fit in with 6, 8 or even 10 or 12 other bells.

Look at Figure 5. When the bells ring in order from the lightest (number 1, called the *treble*) to the heaviest (number 4, called the *tenor*), they are said to be ringing *rounds*. When the order changes, then the ringers are ringing *changes*. The figure shows the simplest of changes, called *plain hunting*.

```
1  2  3  4
2  1  4  3
2  4  1  3
4  2  3  1
4  3  2  1
3  4  1  2
3  1  4  2
1  3  2  4
1  2  3  4
```

Figure 5 The changes for ringing 'plain hunting'.

Topic 4 Exercises

1 Which element in Group I of the Periodic Table reacts most violently with water?

2 Which element in Group VII is a good bleaching agent?

3 Why are the alkali metals kept under oil?

4 Why are the Group I elements called alkali metals?

5 What happens to iodine crystals when they are heated?

6 Why are objects like car bumpers chromium-plated?

7 What colour is iron(II) sulphate?

8 Which acid is represented by the formula H_2SO_4?

9 What is meant by dilute hydrochloric acid?

10 Why do stinging nettles sting?

11 What is litmus and how does it tell you whether or not a substance is an acid?

12 Complete this word equation:

acid + metal →

13 Name three metals that will not react with dilute sulphuric acid.

14 Name the compounds that are formed when zinc reacts with dilute hydrochloric acid.

15 Explain why hydrogen usually explodes with a pop when it is lit, rather than just burning.

16 Use the index to find the name of the element that reacts with hydrogen to make ammonia.

17 Complete this word equation:

acid + carbonate →

18 Why does a lighted splint go out when it is plunged into a test tube of carbon dioxide?

19 Why is a carbon dioxide fire extinguisher the best one to use on electrical fires? Why would it be dangerous to use water?

20 Explain why lime water can be used to detect carbon dioxide.

21 What is dry ice? What is it used for?

22 Write down the names and formulae of two bases and two alkalis.

23 How is soap made?

24 Why is universal indicator called 'universal'?

25 The pH numbers of three solutions are 2, 7, and 8. Which solution is neutral, and which is the strong acid? Which one might be produced by dissolving indigestion tablets in water?

26 Give the names of three salts that can be found in the home.

27 Cooked blackcurrants are red, but if you pour evaporated milk on them, the juice often turns a blue-green colour. Try to explain why.

28 For each pair of reactants, write down the equation for the reaction and the name of the salt that is formed:
a) calcium and hydrochloric acid
b) sodium carbonate and sulphuric acid
c) magnesium oxide and nitric acid.

29 Which two types of chemicals are needed for a neutralisation reaction to take place?

30 Describe how you would make crystals of zinc sulphate, starting with pieces of zinc metal.

31 Why are:
a) cars made of steel (iron)?
b) wires made of copper?
c) bells made of bronze?

32 You are given rods made of different metals and non-metals, all the same length and thickness. You also have some wax, and the usual laboratory apparatus. See if you can design an experiment to compare how well metals and non-metals conduct heat. Describe the experiment.

33 Aluminium left in the air soon goes dull, and new pennies quickly go brown, but gold is always shiny. Explain why this is so.

34 Why would it be dangerous to try putting out a fire of burning magnesium, using water?

35 What information does the Activity Table tell you?

Things to do

36 This is an experiment for you to try, under the supervision of your teacher. The aim is to find out what conditions are needed to prevent iron from rusting.

Put iron nails into test tubes, under these different conditions:
a) nail exposed to air and water
b) nail in tap water but not air
c) nail in a stoppered test tube with a few pieces of calcium chloride. (The calcium chloride will absorb any moisture and keep the air dry.)
d) nail in a stoppered test tube of air-free water. (Freshly boiled water has no air in it.)
e) nail coated with oil
f) painted nail, standing in tap water
g) nail attached to a piece of magnesium ribbon, standing in tap water.

Look at your results and discuss them with your teacher.

Materials for Industry

Crude oil is broken down in a refinery to supply many important materials for industry.

5.1 Metals and man

In history lessons you read about the Stone Age, the Bronze Age and then the Iron Age. To find out why they came in that order, read on . . .

Ancient man

One important thing that distinguished early humans from the apes was their ability to use stones and bits of wood as tools. They used sharp stones for cutting, and hollow stones as cups and dishes. Then they found that a certain type of stone – *flint* – could be chipped and broken into different shapes for arrow heads and knives. See Figure 1.

The first metals that early humans came across were gold, silver and copper. They found them in stream beds and rivers, washed out of the rocks. They are called *native* metals, because they are unreactive enough to be found in the ground in the pure state. The more reactive metals are found combined with non-metals such as oxygen and sulphur, in rocks called *ores*. However, a very small amount of unreacted iron has been found in the form of meteorites. The North American Indians discovered iron in this form, long before we knew how to obtain it from its ore.

The first discovery of metals may have happened 8000 years before the birth of Christ (8000 BC) but another 2000 years passed before man learned how to smelt copper and tin from their ores. He probably found out accidentally, when he was using stones to prop up a fire. If the stones contained copper and tin compounds, the metals would flow out as liquids and quickly harden to lumps.

The Bronze Age started when tin and copper were mixed – again probably by accident – to make a much harder metal. This was an *alloy* called *bronze*. (An alloy is a mixture of different metals.) Bronze could be made of varying hardness, by mixing different proportions of copper and tin. Because of this, it soon became the most important metal of that time. (See Figure 2.) But finding the tin and copper to make it was a problem. Mining was too difficult and dangerous. One technique was to light a fire under metal-bearing rock and then, when the rock was hot, to pour cold water over it and make it crack open. Then man learned that the metals could be extracted much more easily from their ores, if charcoal was mixed with the ores before heating.

As man's understanding increased, so did his technology. For example he learned how to blow air into a furnace with bellows to make the fire hotter. And so, by the year 2000 BC, he could make fire hot enough to smelt iron from iron ore. He had moved into the Iron Age.

At first the iron was very impure and weak. But it improved as different ways were developed to make the furnaces hotter. And soon man found that a little charcoal mixed with the molten iron made a much much stronger metal. The Iron Age had turned to the Steel Age.

Figure 1 Early humans relied on stone for their tools and weapons. The technology needed to extract metals from their ores was only invented relatively recently.

Figure 2 Humans have used metals for decoration and display for thousands of years. Compare this early bronze ornament with the modern jewellery made from gold and diamonds.

Modern man

Between 1750 and 1850 there were rapid advances in transport and building and machinery. The changes were so great that historians call this period the *Industrial Revolution*. For example steam engines and railways were developed. Canals and canal boats were built. Power looms and spinning machines were invented. See Figure 3.

These advances all called out for more and better iron and steel. At the beginning of the 18th century, iron was still smelted from its ore using charcoal, as it had been since the Iron Age. The technology for doing it had improved, however. Furnaces were now tall towers through which hot air was blasted to make the ore and charcoal white hot. They were called *blast furnaces*. But the iron they produced was often very impure and weak. Then, in 1777, a breakthrough was made. Abraham Darby, an iron maker, hit on a better way of extracting iron from its ore. Instead of using charcoal he used *coke*, which he made by roasting coal until only its carbon skeleton was left.

The iron made by this method was so much better that it could be used with confidence for the frames of machines, and for rails on the new railways, and even for ships and bridges. As a monument to this great change, Abraham Darby's grandson built an iron bridge over the river Severn, in Shropshire, at the place that we now call Ironbridge. (See photo on page 36.)

Very modern man

Of course, iron and steel are not the only metals we use today. Titanium, tungsten, aluminium, copper, zinc and many others are very important in the modern world. But iron is still the most widely used metal of all. It is always mixed with other metals, a whole range of steels. For example tungsten makes very hard steel for drill tips, while cobalt makes steel ideal for electromagnets. Chromium and nickel give *stainless steel* for car parts and cutlery and sinks.

The story of metals is not complete, however. We still need to develop new materials to cope with new problems. Here are just a few that are needed: alloys that can withstand the high temperatures and radiation in nuclear reactors; new steels for North Sea oil rigs, that will survive very low temperatures and battering by heavy seas; new light alloys for aeroplanes; metals for satellites; metals for semiconductors in electronic components (see Figure 4); carbon fibre alloys for turbine blades in power stations and jet engines. . . . The list could go on and on.

THE ROCKET.

Figure 3 During the Industrial Revolution, there were rapid advances in transport and machinery. These advances included the invention of the steam engine and of power machines in the textile industry.

Figure 4 Modern advances in metal technology have led to the development of the miniature silicon chip. For further advances to be made, new metals are needed.

Exercises

1 What is a native metal?

2 What is an alloy? Give an example.

3 Why did the Bronze Age come before the Iron Age?

4 Give an example of something that was first made during the Industrial Revolution, and that needed iron or steel for its manufacture.

5 What was different about Abraham Darby's iron?

6 Try to think of another example of a modern-day development, that needs a material that can stand up to extreme conditions.

5.2 The extraction of metals

The first metals to be discovered were those that occurred in nature as pure elements, like gold and silver. Others had to wait a lot longer, especially if they were difficult to extract.

Native metals

The *native metals* are those that are found in the ground uncombined with other elements. They are the unreactive metals, that come at the bottom of the Activity Series. (See Figure 1.) One example is gold. Most of the world's gold comes from South Africa – about 900 tonnes of it a year. Rock containing small traces of the metal is mined from thousands of metres below the surface of the Earth. Next it is crushed and the gold washed out. The gold is then melted, and poured into moulds to make *ingots*. An ingot of gold is shown in Figure 2.

Roasting ores

Metals like lead, copper, and mercury occur in the ground as compounds. But they are not strongly combined in these compounds, since they are not very reactive metals. They can be extracted by roasting the ores in air in a furnace.

Most of our copper comes from the USA, Canada, and Zambia. It occurs mainly as *chalcopyrite*, $CuFeS_2$. To extract the copper, the ore is first crushed and mixed with a frothy oil–water mixture. The bits that contain copper float to the top, in the froth. They are then separated from it, and roasted in a furnace with limestone while air is blown in. (The limestone is to remove impurities.) This reaction takes place:

copper sulphide + oxygen → copper + sulphur dioxide gas

The molten copper is cast into ingots. The limestone combines with impurities and makes a waste product called *slag*.

The copper obtained in this way is not at all pure. But it can be made 99.9% pure by a process called *electrolysis*. Look at Figure 3.

potassium	K
sodium	Na
calcium	Ca
magnesium	Mg
aluminium	Al
zinc	Zn
iron	Fe
lead	Pb
copper	Cu
silver	Ag
gold	Au

decreasing reactivity

Figure 1 Silver and gold are very unreactive metals. They are placed at the bottom of the Activity Series.

Figure 2 After gold has been extracted from its ore, it is melted and poured into a mould to make an ingot. This ingot has a mass of 12.5 kg.

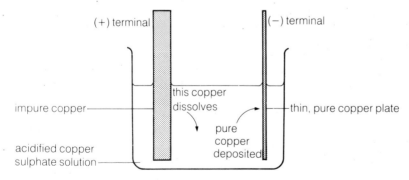

(+) terminal   (−) terminal

impure copper —

this copper dissolves

pure copper deposited

thin, pure copper plate

acidified copper sulphate solution —

Figure 3 Copper is purified by electrolysis.

The impure copper is made into a thick plate which is then connected to the positive terminal of a battery. A thin plate of pure copper is connected to the negative terminal. Both plates are put in a bath of acidified copper sulphate solution. As the current flows through this solution, the impure copper dissolves and pure copper is deposited on the thin plate.

Copper is very important for making electric wires, water pipes, and alloys like brass and bronze.

Reducing ores

More reactive metals like iron and zinc need more powerful methods of extraction from their ores.

Iron is found in the ground as *haematite* (Fe_2O_3), *magnetite* (Fe_3O_4), and *siderite* ($FeCO_3$). An ore like haematite must be *reduced* to get the iron out. That means the oxygen must be removed from it. (The removal of oxygen from a compound is called *reduction*.) This is carried out with the help of carbon, in the form of *coke* which is made from coal.

Look at Figure 4. The reduction takes place inside a *blast furnace*. Iron ore, coke, and limestone are put in at the top. Very hot air is blown in at the bottom. Inside, these reactions take place:

1 The carbon reacts with the air to make carbon monoxide:

$$C + O_2 \rightarrow CO_2 \text{ and then } CO_2 + C \rightarrow 2CO$$
carbon monoxide

2 The carbon monoxide reduces the iron ore:

$$3CO + Fe_2O_3 \rightarrow 3CO_2 + 2Fe$$

The molten iron trickles down to the bottom of the furnace.

3 The limestone reacts with sand and other impurities in the iron ore to form a molten substance called slag. This also runs to the bottom where it floats on the iron.

The molten iron has up to 4% carbon dissolved in it. Some of it is run into moulds, where it hardens to form *cast iron*. Because of the carbon, cast iron is hard but brittle – it breaks easily. So it is used only for things that do not have to bear great loads, such as lamp posts and iron railings. (See Figure 5.)

Most of the molten iron is carried away to an *oxygen converter*, where it is made into *steel*. In the converter it is mixed with scrap iron and more limestone, and strongly heated. Oxygen is blown onto the molten mixture, and most of the carbon reacts with it to form carbon dioxide. The limestone turns any other impurities into slag.

The steel that comes out of the converter contains only 0.3% to 1.5% carbon, and is strong enough to be made into large things like bridges and building girders, as well as small things like screwdrivers and spanners. Often, other metals are also mixed in with the steel to make different types of *alloy steel*. Look at Figure 6.

Electrolysis

Very reactive metals like aluminium, sodium, and magnesium are so strongly combined with other elements in their ores that they can only be removed using electricity.

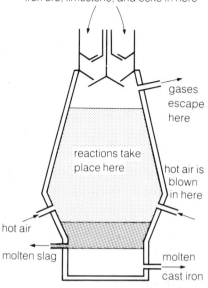

iron ore, limestone, and coke in here

gases escape here

reactions take place here

hot air is blown in here

hot air

molten slag

molten cast iron

Figure 4 Iron ore is reduced inside a blast furnace. Molten iron is produced, together with slag and waste gases.

Figure 5 These railings are made from cast iron.

alloy steel	used for
manganese steel	railway points
tungsten steel	tools
chromium steel	ball bearings
nickel/chromium steel	stainless steel
molybdenum steel	propeller shafts, rifle barrels
cobalt steel	magnets

Figure 6 An alloy steel is a mixture of steel with another metal. Different alloy steels are used for different purposes.

Aluminium occurs as aluminium oxide or **bauxite**, Al_2O_3. This comes from places like Australia and the USA. To obtain aluminium, the oxide is dissolved in a molten substance called **cryolite**. Then the mixture is put into an electrolysis tank. Look at Figure 7. When the current flows, oxygen is formed at the carbon electrode and molten aluminium runs to the bottom of the tank. From there it is siphoned off.

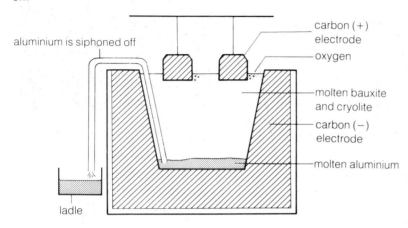

aluminium is siphoned off

carbon (+) electrode

oxygen

molten bauxite and cryolite

carbon (−) electrode

molten aluminium

ladle

Figure 7   Aluminium is extracted from bauxite by electrolysis.

Because it is such a light metal, aluminium is used for making aeroplanes, buses, and underground trains. It can be squeezed into different shapes when hot, so it is made into window frames and greenhouse structures as well. It can also be rolled into very thin sheets to make cooking foil. Look at Figure 8.

Figure 8   Kitchen foil is thin aluminium.

Exercises

1 Arrange these metals in order of activity, with the most reactive one at the top:

   copper   potassium   aluminium   gold   iron

2 What name is given to the rock from which a metal is extracted?

3 Where does most of the world's supply of gold come from?

4 Give one important use for copper.

5 Name one ore of iron.

6 What does 'reduction' mean?

7 Name one metal that is extracted from its ore by electrolysis.

5.3 Limestone

More than 2000 years ago, the Romans used limestone for building and agriculture. Today, we still put it to the same uses.

How much, and what for?

Somewhere near you there is probably a limestone quarry, at the side of a hill made of limestone. Every day, part of the hillside is blasted with explosives and scooped up with mechanical diggers. Look at Figure 1. Lorries carry the huge pieces of rock to crushers and from there, the rock is sent to various industries.

Most limestone goes to the steel industry, where it is used in the production of iron and steel. In the furnaces, it reacts with impurities in the iron and steel, turning them into *slag* that can easily be removed.

A large amount of limestone is also used for making other chemicals, such as sodium carbonate for the glass industry. Smaller amounts are used for building, water treatment, agriculture, paper making, and sugar refining. In all, about 4 million tonnes of limestone are used in Britain each year.

The lime kiln

Limestone is often roasted in a *lime kiln* to produce calcium oxide or *quicklime*. Look at Figure 2. Brick-sized pieces of limestone are packed into the kiln and hot gas flames are injected through the sides. The limestone decomposes to give calcium oxide and carbon dioxide:

$$CaCO_3 \rightarrow CaO + CO_2$$

limestone   quicklime   carbon dioxide

As the calcium oxide falls to the bottom of the kiln it is removed, and more limestone moves down to take its place.

Many old lime kilns can be seen around the country. They were heated by wood or coal and the production of quicklime took many days. An engraving of old lime kilns in the Wye Valley is shown in Figure 3.

Slaked lime

Much of the quicklime from the lime kiln is added to water, to produce calcium hydroxide or *slaked lime*:

$$CaO + H_2O \rightarrow Ca(OH)_2$$

quicklime   slaked lime

This is a highly exothermic reaction. The quicklime cracks and steams, and with excess water forms a white suspension called *milk of lime*.

If the milk of lime is dried and powdered it can be mixed with sand and water to make *mortar* for building. It can also be spread on farm land to neutralise the acidity of the soil.

Figure 1 This digger is removing rock from a limestone quarry.

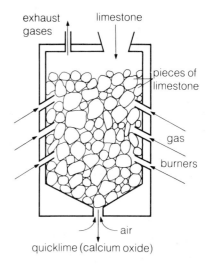

Figure 2 Limestone is roasted in a lime kiln to produce quicklime.

Figure 3 Lime kilns as they used to be

Lime water

A saturated solution of calcium hydroxide is called *lime water*. It is a weak alkali. Carbon dioxide is an acidic gas. When carbon dioxide is bubbled through lime water, the acid neutralises the alkali and a salt is formed:

$$CO_2 + Ca(OH)_2 \rightarrow CaCO_3 + H_2O$$

acid · · · · alkali · · · · salt · · · · water

The salt is calcium carbonate, which is insoluble. It is formed as a white suspension which makes the lime water cloudy.

The lime water test for carbon dioxide is a very sensitive one. If the lime water goes cloudy, then the gas must be carbon dioxide. Look at Figure 4.

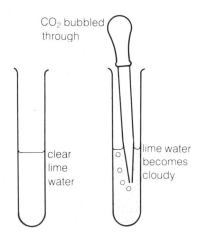

Figure 4 If a gas is bubbled through lime water and the lime water goes cloudy, then the gas is carbon dioxide. This is called the lime water test for carbon dioxide.

Cement

Mortar, made from slaked lime and sand, crumbles after a few years of exposure to the air. *Cement* is much longer-lasting.

Ground-up limestone is mixed with clay and water to form a runny mixture called a *slurry*. The slurry is poured into a rotating kiln. Look at Figure 5.

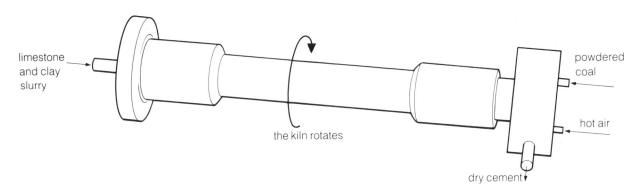

Figure 5 A cement kiln

Powdered coal is injected into the other end of the kiln. As the coal dust burns, the slurry dries out to form a mixture of calcium and magnesium silicates. This complicated mixture is mixed with *gypsum* (calcium sulphate) and it becomes Portland cement.

When cement is mixed with sand and water and left to dry, a strong crystal lattice is formed that can bind bricks together. If gravel or small stones are added to the cement, *concrete* is formed, and that is even stronger. *Reinforced concrete* has metal bars and steel mesh embedded in it. This makes it strong enough for such things as bridges and buildings. Look at Figure 6.

Limestone for building

Limestone is a good building material. It is strong enough to withstand normal attack by wind and rain, and soft enough to be cut into blocks and chiselled into shapes. Moreover, weathered limestone looks attractive. However, limestone is a carbonate and carbonates are dissolved by acids. In industrial areas, sulphur

Figure 6 This concrete has been reinforced with steel mesh.

dioxide escapes into the atmosphere from factory chimneys. Most of it is carried high into the air, but there it dissolves in rain, forming a very dilute solution of sulphuric acid. This falls as *acid rain*, and slowly dissolves the limestone. The erosion is helped on by fumes from car exhaust, which contain *nitrogen dioxide*, another acidic gas. Figure 7 shows part of a limestone building that has been badly eroded.

Besides being a problem in cities, acid rain poses a very serious threat to plants. Normal rain has a pH value of about 5 or 5.5. Acid rain can have a pH number of 4 or less. In parts of the Black Forest in Germany, some lakes and ponds have been found to have a pH as low as 3: 'as acid as lemon juice', as one newspaper put it. This acidity slowly kills trees, plants, and fish. Countries like Sweden seem to get more acid rain than other places, probably because of the heavily industrialised countries round about them. Almost certainly in the very near future, in an attempt to stop acid rain, we shall see laws passed to control the sort of pollution that factory chimneys are allowed to give out.

Figure 7 This building has been eroded by acid rain.

Exercises

1 Name three industries that use limestone.

2 What are the chemical names and formulae of limestone, quicklime, and slaked lime?

3 What are the starting materials for cement?

4 Explain the difference between cement and concrete.

5 Describe how you would use lime water to test a gas, to see if it was carbon dioxide.

5.4 Sulphuric acid

In the past, chemists made sulphuric acid by heating crystals of zinc sulphate, which they called white vitriol. The oily liquid they obtained was called oil of vitriol.

Some uses of sulphuric acid

A modern industrial country must have sulphuric acid. In Great Britain, about 3 million tonnes of the chemical are made each year. What is it all used for? Look at Figure 1 to find out.

In order to grow all the food we need, farmers must put fertilizers on the soil. Two that are commonly used are *ammonium sulphate* and *superphosphate* (a mixture of calcium phosphate and calcium sulphate). Sulphuric acid is needed to make both.

When the farmer paints his house, the white paint he uses will probably contain *titanium dioxide*. This white pigment is extracted from its ore using sulphuric acid. Look at Figure 2.

Thousands of gallons of *detergent* are used each week, just for washing up. Detergent is made by reacting chemicals from crude oil with sulphuric acid.

Steel objects that are going to be electroplated with nickel and chromium to make them shiny (such as electric kettles and kitchen taps), have to be thoroughly cleaned first with dilute sulphuric acid. This process is called *pickling*.

And finally, one of its miscellaneous uses is for car batteries (see Figure 3). Think of all the dilute sulphuric acid that must go into all the car batteries sold in shops and garages.

The Contact process

Since the beginning of the century, sulphuric acid has been made by the *Contact process*. There are three important parts to it:

1 Preparing the starting materials These are sulphur dioxide and oxygen. The sulphur dioxide is made by burning sulphur in air. The sulphur reacts with the oxygen in air:

$$S + O_2 \rightarrow SO_2$$
sulphur dioxide

2 The reaction The sulphur dioxide is made to react with more oxygen, forming a gas called sulphur trioxide:

$$2SO_2 + O_2 \rightleftharpoons 2SO_3$$
sulphur trioxide

Look at the sign in the middle of the equation. It shows that the reaction is an *equilibrium reaction*. This means that not all the sulphur dioxide and oxygen will react together. To get as much as possible to react (about 98% in fact), certain steps are taken:

i) The gases are heated to 450 °C before they react.

ii) The gases are compressed to about twice atmospheric pressure.

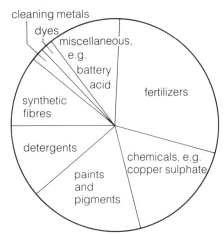

Figure 1 About 3 million tonnes of sulphuric acid are made in Britain each year. This pie chart shows where it all goes.

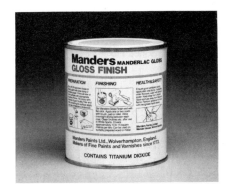

Figure 2 The white pigment in paint is titanium dioxide. Sulphuric acid is used to extract the pigment from its ore.

Figure 3 The liquid in a car battery is sulphuric acid. (Note the corrosive warning label.)

iii) A *catalyst* of vanadium(V) oxide is used. (See Figure 4.) A catalyst is something that changes the rate of a reaction. Here it is used to speed up the reaction.

3 Absorbing the sulphur trioxide The sulphur trioxide gas is passed into concentrated sulphuric acid. It reacts with the acid to give a thick, oily liquid called *oleum*. This is very carefully diluted with water to make ordinary concentrated sulphuric acid.

Figure 5 shows the three parts of the process in a simple flow diagram.

Figure 4 Vanadium(V) oxide catalyst pellets. These pellets are used to speed up the reaction of sulphur dioxide with oxygen, in the Contact process.

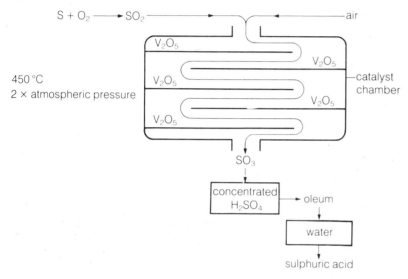

Figure 5 The Contact process for making sulphuric acid

Sulphuric acid is expensive. It costs more than £6 million to build a Contact process plant, and making 100 g of sulphuric acid uses as much energy as burning 10 000 one-hundred-watt light bulbs for one hour. However, nearly three-quarters of the heat used is reclaimed, and sold as hot steam to other nearby industries.

A sulphuric acid plant

Sulphuric acid in the laboratory

Sulphuric acid takes part in the usual reactions of acids, which you met in Unit 4.1.

1 Sulphuric acid reacts with many metals to form hydrogen gas.
2 Sulphuric acid reacts with alkalis and neutralises them.
3 Sulphuric acid reacts with carbonates and produces carbon dioxide.

For all these reactions, *dilute* sulphuric acid is used. When the acid is *concentrated*, it has quite different properties. It is then a *dehydrating agent*. That means it will remove water from other substances.

Example 1

'Wet' gases are bubbled through concentrated sulphuric acid to dry them. Look at Figure 6.

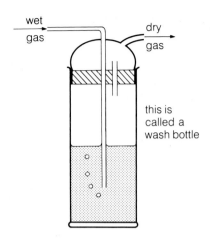

Figure 6 Concentrated sulphuric acid is used to dry gases.

Example 2

Concentrated sulphuric acid will take the water out of sugar, leaving carbon behind. Look at the equation:

$$C_{12}H_{22}O_{11} \quad - \quad 11H_2O \quad \rightarrow \quad 12C$$

sucrose (sugar)   black carbon

Example 3

Concentrated sulphuric acid will also dehydrate your skin if it gets on it. The reaction can cause very bad blisters, since it gives out a great deal of heat. So, if you spill concentrated sulphuric acid on your skin, *wash it off at once* with lots and lots of water, and tell your teacher immediately.

Making dilute sulphuric acid

When concentrated sulphuric acid is mixed with water to make ordinary laboratory acid, it gets very hot indeed. It gets so hot that it may boil up and spit out of the beaker onto your hands and into your eyes. So, when you are diluting concentrated sulphuric acid, follow these rules:

1 Wear gloves and a face mask.
2 *Always* add *acid* to *water*, *slowly*.
3 *Never* add water to acid, as it gets too hot this way.

Look at Figure 7.

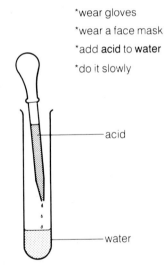

Figure 7 Always follow these rules when diluting concentrated sulphuric acid.

Exercises

1 Give three uses of sulphuric acid.
2 What is 'pickling'?
3 What are the starting materials in the Contact process?
4 Write a balanced equation for the reaction of sulphur with oxygen.
5 What three steps are taken in order to produce as much sulphur trioxide as possible during the Contact process?
6 Describe how you would dilute a sample of concentrated sulphuric acid.

5.5 Ammonia and nitric acid

When you breathe in air to get oxygen, you breathe the nitrogen out again unused. But that isn't the end of it. It probably finishes up in your vegetables!

Figure 1 Lightning turns rain water into dilute nitric acid. Such rain is naturally acidic and is essential for the soil and for the growth of plants. Pollution chemicals such as sulphur dioxide and nitrogen dioxide get into the air and make it more, unnaturally acidic. This acid rain is dangerous because it can kill plants and trees.

Nitrogen in the air

A flash of lightning contains a huge amount of energy. During a thunderstorm, the lightning can split oxygen and nitrogen molecules in the air, into their separate atoms. The oxygen and nitrogen atoms then combine to make *oxides of nitrogen*, and these dissolve in the rain and fall to earth. In the soil, they react with the soil chemicals to form solutions of *nitrates*. Nitrates are salts of nitric acid, so you could say that it rains dilute nitric acid! Look at Figure 1.

Of course the acid is very dilute, but it plays a very important part in the growth of plants. Plants take in the nitrates through their roots, and use them to make substances called *proteins*. Then humans and other animals take in the proteins when they eat the plants. Without proteins none of us could live.

Nitric acid in the laboratory

Dilute nitric acid is just like other laboratory acids:

1 It turns litmus paper red.

2 It neutralises bases. For example:

$$\text{copper(II) oxide} + \text{nitric acid} \rightarrow \text{copper(II) nitrate} + \text{water}$$
$$CuO + 2HNO_3 \rightarrow Cu(NO_3)_2 + H_2O$$

3 It reacts with carbonates. For example:

$$\text{calcium carbonate} + \text{nitric acid} \rightarrow \text{calcium nitrate} + \text{carbon dioxide} + \text{water}$$
$$CaCO_3 + 2HNO_3 \rightarrow Ca(NO_3)_2 + CO_2 + H_2O$$

But dilute nitric acid is different from the other acids in the way it reacts with *metals*. Instead of forming hydrogen, it *oxidises* the metal and forms nitrogen oxide instead. What's more, it reacts with copper which other dilute acids will not do:

$$\text{copper} + \text{nitric acid} \rightarrow \text{copper(II) nitrate} + \text{water} + \text{nitrogen (II) oxide}$$
$$3Cu + 8HNO_3 \rightarrow 3Cu(NO_3)_2 + 4H_2O + 2NO$$

The Haber process

Earlier this century, scientists in Norway tried to imitate the effect of lightning. They tried to make nitric acid by blowing oxygen and nitrogen through a huge spark. They succeeded, but the method was too costly to be much use. Today, we make the acid from *ammonia*. But first the ammonia is made by the *Haber process*.

This process was devised by a German scientist during the First World War, when the Germans badly needed ammonia to make explosives and fertilizers. See Figure 2.

Figure 2 The Haber process is named after Fritz Haber. He was a German scientist and he devised the process during the First World War. It is still used today.

Like the Contact process, the Haber process has three important stages:

1 Preparing the starting materials These are nitrogen (from the air), and hydrogen (made from natural gas, CH_4).

2 The reaction This is the equation:

$$N_2 + 3H_2 \rightleftharpoons 2NH_3$$
ammonia

It is an equilibrium reaction. To get as much ammonia from it as possible, three steps are taken:

i) The gases are heated to a temperature of 450 °C.

ii) They are compressed to 200 times atmospheric pressure.

iii) A catalyst of iron is used.

3 Isolating the ammonia When the ammonia and unreacted nitrogen and hydrogen come out of the reaction chamber, the ammonia is cooled so that it liquefies. Figure 3 shows a simple flow diagram for the process.

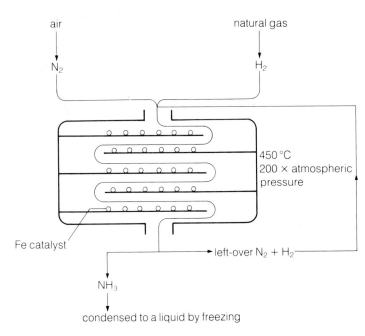

air

natural gas

N_2

H_2

450 °C
200 × atmospheric pressure

Fe catalyst

left-over $N_2 + H_2$

NH_3

condensed to a liquid by freezing

Figure 3 The Haber process for the manufacture of ammonia

Altogether, British factories make about 7000 tonnes of ammonia a day. Throughout the world, more than 50 million tonnes are made each year. Figure 4 shows some of its uses.

Making nitric acid

To make nitric acid, the ammonia is mixed with air and passed over a hot platinum catalyst. Oxides of nitrogen are formed and these are dissolved in water to give nitric acid.

Nitric acid is used to make even more fertilizers, and explosives like TNT (trinitrotoluene) and nitroglycerine.

Figure 4 Ammonia is used in the manufacture of nitric acid, explosives, fertilizer, fabric dyes, and synthetic fibres.

Ammonia in the laboratory

Ammonia gas is very soluble in water, and forms a solution called *ammonium hydroxide*, or *ammonia solution*:

$NH_3 + H_2O \rightarrow NH_4OH$
ammonia solution

Ammonia solution is an alkali. It turns litmus paper blue and it neutralises acids, forming salts. For example:

ammonia solution	+	sulphuric acid	$\rightarrow$	ammonium sulphate	+ water
$2NH_4OH$	+	H_2SO_4	$\rightarrow$	$(NH_4)_2SO_4$	+ $2H_2O$

Ammonium salts are important. Ammonium sulphate and ammonium nitrate are used as fertilizers, and ammonium chloride is one of the chemicals inside dry-cell batteries.

Ammonia solution is very good at dissolving grease. Because of this, kitchen cleaning agents often contain 'powerful ammonia'. Look at Figure 5. However, you must take care when using them because they give off ammonia and it is a pungent, choking gas.

Figure 5 Kitchen cleaning agents often contain 'powerful ammonia'. Ammonia solution is very good at dissolving grease.

Exercises

1 How does lightning play an important part in the growth of plants?

2 Briefly explain how ammonia is made into nitric acid.

3 In what way is dilute nitric acid different from other dilute acids?

4 What are the starting materials in the Haber process?

5 What conditions of temperature and pressure are used in the Haber process, and what is the catalyst?

6 Why is ammonia such an important chemical?

5.6 Electrolysis (1)

You saw earlier that a process called electrolysis is used to extract certain metals from their ores. We will now take a closer look at this process.

Conductors and insulators

You are probably quite used to the idea of solids conducting electricity. Metals like copper, aluminium, zinc, iron, and magnesium allow electricity to pass through them easily. In fact all metals will conduct electricity, and so will the non-metal graphite (a form of carbon). They are called *conductors*.

Substances like plastic, glass, and rubber do not conduct electricity, and they are called *non-conductors* or *insulators*. Look at Figure 1. In an electric plug, rubber and plastic are used as insulators to keep the conductors (the copper wires) from touching each other.

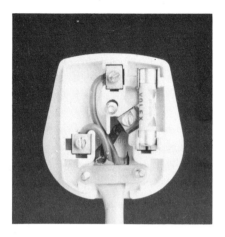

Figure 1 In an electric plug, insulators (rubber and plastic) stop the conductors (copper wires) from touching each other.

Electricity and liquids

Some liquids conduct electricity too. You should know that it is dangerous to touch anything electrical with wet hands, because you might get an electric shock. Liquids that conduct electricity are called *electrolytes*. They are usually solutions of acids or alkalis, solutions of salts, or molten salts. For example, dilute sulphuric acid, sodium hydroxide solution, sodium chloride solution, and copper sulphate solution are all electrolytes. So are molten lead bromide, molten sodium chloride and molten potassium iodide.

Liquids that do not conduct electricity are called *non-electrolytes*. Examples are ethanol, petrol, sugar solution, and molten wax.

Whenever electricity flows through an electrolyte, the liquid decomposes in some way. This is called *electrolysis*. You will see an example of electrolysis in a moment.

Apparatus for electrolysis

If you want to pass electricity through a liquid, you first need a source of electricity – such as a battery. Next you need some way of getting the electricity into the liquid. For this you can use *electrodes*. These are rods or strips of graphite or of metals like platinum, gold, or silver. These materials are chosen because they are conductors, and because they are very unreactive, so will not react with the liquids you put them in.

Look at Figure 2. Notice that the electrodes have special names. The electrode connected to the positive terminal of the battery is called the *anode*. The electrode connected to the negative terminal of the battery is called the *cathode*.

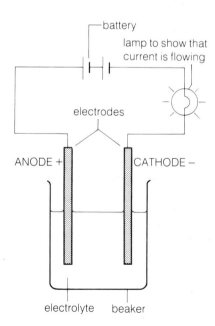

Figure 2 The apparatus for electrolysis includes a battery, electrodes, and an electrolyte.

What exactly goes on?

All substances are made of atoms or groups of atoms called molecules. In electrolytes, the atoms and molecules have positive or negative charges on them. These charged atoms and molecules are called *ions*. Electrolytes contain ions. Non-electrolytes do not contain ions.

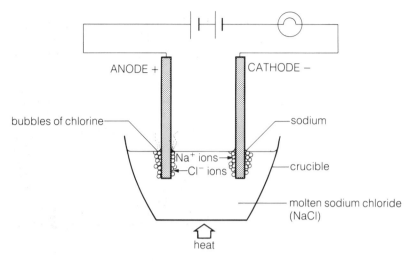

Figure 3 The electrolysis of molten sodium chloride produces sodium and chlorine.

Figure 4 A Down's cell is used to make sodium and chlorine from molten sodium chloride.

Look at Figure 3. The liquid in the crucible is molten sodium chloride. Sodium chloride contains positively-charged *sodium ions* (Na^+) and negatively-charged *chloride ions* (Cl^-). When the electrodes are dipped into the electrolyte and the electricity is switched on, the positively-charged sodium ions are attracted towards the negatively-charged electrode (the cathode) since it has the opposite charge. When they get there they change into sodium. In the same way the chloride ions are attracted to the anode and change into chlorine gas. This process is called *electrolysis*. So, during the electrolysis of molten sodium chloride, liquid sodium is formed at the cathode and chlorine is formed at the anode. The sodium chloride has decomposed to sodium and chlorine.

Figure 4 shows a Down's cell for making sodium and chlorine from molten sodium chloride. Millions of tonnes of these chemicals are made each year in this way. In another apparatus called a Flowing Mercury cell, sodium hydroxide and chlorine are made from sodium chloride solution. Look at Figure 5.

Figure 5 Flowing Mercury cells are used to make sodium hydroxide and chlorine from sodium chloride solution.

Exercises

1 Give two examples of conductors and two of non-conductors.

2 What name is given to a liquid that conducts electricity?

3 Name a liquid that will not conduct electricity.

4 What are electrodes generally made of? Explain why.

5 What are the other names for the positive and negative electrodes?

6 What are ions?

7 If molten magnesium chloride is used in an electrolysis experiment, what substance would you expect to be formed at: **a)** the anode? **b)** the cathode? Explain why.

r

5.7 Electrolysis (2)

Now that you've got the idea of electrolysis, you can find out more about it below . . .

Are there any rules?

Yes, there are.

1 Electrolytes are made up of charged atoms or groups of atoms. These are called *ions*.

2 Metal atoms in electrolytes are positive ions. Hydrogen ions are also positive.

3 Non-metal atoms like chlorine and bromine, and groups of atoms like hydroxide, sulphate, and nitrate, are all negative ions.

4 Positive ions are always attracted towards the negative electrode, the cathode. (Because of this they are called *cations*.)

5 Negative ions are always attracted towards the positive electrode, the anode. (Because of this they are called *anions*.)

Electrolysis is used to coat bicycle handlebars with chromium.

The rules in action

Look at these examples.

1 **The electrolysis of molten lead bromide.** Look at Figure 1. The lead bromide is made of positive *lead ions* (Pb^{2+}) and negative *bromide ions* (Br^-). The lead ions are attracted to the cathode and turn into molten lead. The bromide ions are attracted to the anode and change into bromine gas.

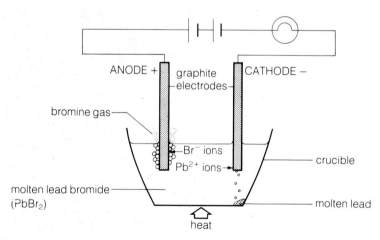

Figure 1 The electrolysis of molten lead bromide produces molten lead and bromine.

2 **The electrolysis of dilute sulphuric acid.** This is a little more difficult, because dilute sulphuric acid is a mixture of sulphuric acid and water. What happens is that the sulphuric acid helps the water to change into *hydrogen ions* (H^+) and *hydroxide ions* (OH^-). At the electrodes, these change into the gases hydrogen and oxygen.

Look at Figure 2. The hydrogen ions are attracted towards the cathode and change into hydrogen gas. The hydroxide ions are

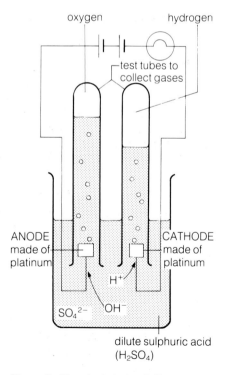

Figure 2 The electrolysis of dilute sulphuric acid produces hydrogen and oxygen. Twice as much hydrogen as oxygen is produced.

attracted towards the anode where they form oxygen gas. The volume of hydrogen that forms is twice the volume of oxygen. This is not surprising when you look at the formula of water: H_2O.

3　**The electrolysis of copper sulphate solution.** Because the copper sulphate is dissolved in water, the solution is a mixture of hydrogen and hydroxide ions from the water, and copper ions (Cu^{2+}) and sulphate ions (SO_4^{2-}) from the copper sulphate. Of these, only the *copper ions* and *hydroxide ions* take part. Look at Figure 3. The copper ions are attracted towards the cathode and change into copper. The cathode becomes coated with brown copper. The hydroxide ions are attracted towards the anode and change into oxygen gas.

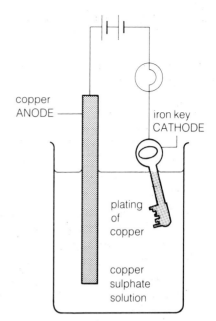

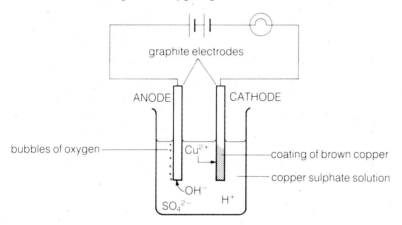

Figure 3　The electrolysis of copper sulphate solution produces copper and oxygen.

Figure 4　In electroplating the metal in the anode dissolves and coats the cathode. In this example, the key becomes coated with copper.

Electroplating

The electrodes used in the examples so far were chosen because they are *inert*. That is, they do not dissolve in or react with the electrolyte. But sometimes this is just what you want to happen. Look at Figure 4. Here the electrolyte is copper sulphate solution, and the anode is made of copper. The cathode is an iron key. When the electricity is switched on, the copper anode slowly dissolves and the cathode (the key) becomes coated or plated with copper. This process is called *electroplating*.

In electroplating, the object to be plated is always used as the cathode. The metal you want to plate it with is made into the anode. It could be silver or gold or nickel or chromium. For example a lot of jewellery is made of cheap metal and then plated with silver or gold. Look at Figure 5.

Figure 5　These objects have all been electroplated.

Exercises

1　During electrolysis, which type of ion is attracted to: a) the anode?　b) the cathode?

2　For the electrolysis of molten zinc iodide, say what substances you would expect to be formed at:
a) the anode　b) the cathode.

3　In electrolysis, what other names are used for the positive and negative ions? Explain why.

4　During the electrolysis of dilute sulphuric acid, 50 cm^3 of oxygen was formed at the anode. What volume of hydrogen was formed at the cathode?

5　You want to electroplate a bracelet with silver. What would you use as the cathode? What would you use for the other electrode? What liquid do you think you would use as the electrolyte?

5.8 Plastics

The word 'plastic' really means 'easy to mould' – like plasticene. But some of the plastics made these days are very hard indeed.

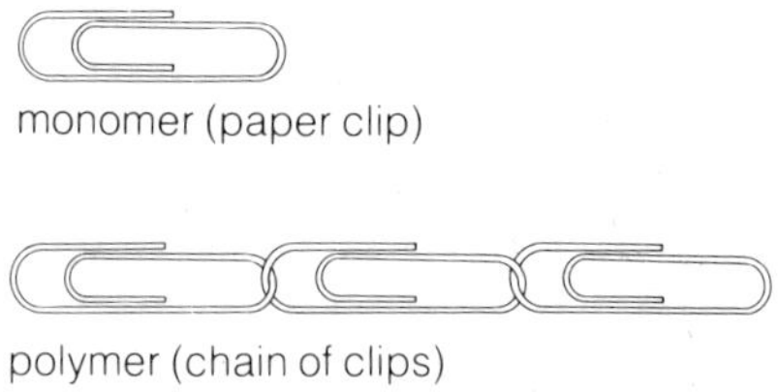

Polymers

Polythene, PVC, polystyrene, terylene, and nylon are all *plastics*. They are made into useful objects by heating them and moulding them into shape. They are also *polymers*.

Polymers are compounds whose molecules are very long chains. These chains are made when a very large number of a small molecule are joined together. The small molecule is called a *monomer*. Look at Figure 1.

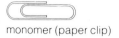

monomer (paper clip)

polymer (chain of clips)

Figure 1 You can picture a polymer as a chain of paper clips. Each paper clip represents a monomer.

Polyethene

Polyethene is often called *polythene* for short. It is a polymer made from the monomer *ethene*. Scientists picture ethene like this:

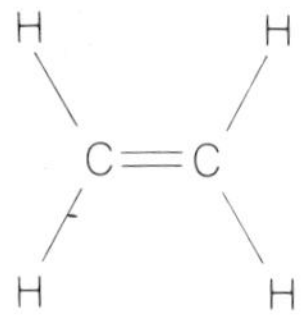

$$H_2C=CH_2$$

The lines drawn between the atoms represent the 'bonds' or forces holding the atoms together. These bonds can be either *single bonds* or *double bonds*. Ethene has a double bond between the carbon atoms.

When ethene gas is heated and compressed over a catalyst, a sticky, white solid is formed, and hardens as it cools. The monomer ethene has *polymerised* and the white solid is called polyethene. Remember that 'poly' means 'many', so 'polyethene' means 'many ethenes'. Look at the equation:

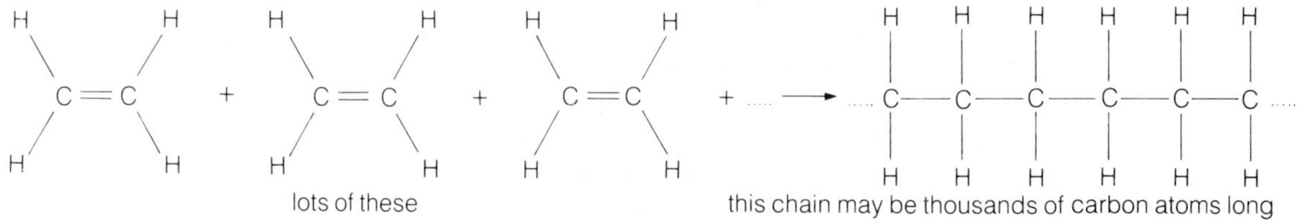

$$H_2C=CH_2 \ + \ H_2C=CH_2 \ + \ H_2C=CH_2 \ + \ \cdots \longrightarrow \ \cdots C-C-C-C-C-C \cdots$$

lots of these   this chain may be thousands of carbon atoms long

Note that the double bonds have turned into single bonds.

Polyethene is used for making washing-up bowls, buckets, toys, dustbins, milk crates, plastic bags, and many other things.

Polystyrene

Polystyrene is a polymer made from the monomer *styrene*. Styrene is a liquid. When it is added to warm water, small beads of polystyrene are formed. The small beads can be melted and moulded into such things as stems for ball-point pens (see Figure 2), model ships and aeroplanes, and imitation glass for holding foodstuffs. If air is blown into the plastic while it is still liquid, *expanded polystyrene* is made. This familiar substance is used for protecting fragile objects, for displaying meat and vegetables in supermarkets (see Figure 3), and for keeping things hot or cold. For example, it is used inside 'cool boxes' to keep the heat out. And some people stick polystyrene tiles on their ceilings to keep heat in.

Figure 2   This pen is made of polystyrene.

Other plastics

Just look at the list of plastics below. Many of them will probably be new to you. Each is made from molecules of a monomer (or sometimes two monomers) joined together to form very long chains.

plastic	uses
polyvinyl chloride (PVC)	raincoats, records, garden hoses and electrical insulation
polyamide (Nylon)	shirts, jumpers, tights
polyester (Terylene)	clothes, seat belts, yacht sails
polypropene	string, carpets
polyurethane	furniture padding
polyvinyl acetate	glue and paint
polytetrafluoroethene ('Teflon')	non-stick surfaces
polymethacrylate (perspex)	safety glass
melamine	children's cups and plates and kitchen shelves
bakelite	electrical fittings

Figure 3   Vegetables in supermarkets are often packaged and displayed on a tray made of expanded polystyrene. The cling film is also a plastic, similar to PVC.

Advantages and disadvantages of plastics

Plastics have many advantages over traditional materials like wood, metal, stone, wool and cotton. They are often cheaper. They are easier to make into complicated shapes. They are resistant to attack by air and water and they are often unaffected by acids and alkalis. In many cases they are lighter and stronger.

However, they have two big disadvantages. First, they do not decompose easily, and this causes a great problem in rubbish disposal. Think how much waste plastic is thrown away. Second, they may give off poisonous gases when they burn. These include carbon monoxide and hydrogen cyanide that form when common furniture plastics like polyurethane catch fire.

Exercises

1   What does the word 'plastic' mean?
2   What is a polymer?
3   What does the 'poly' mean in 'polymer'?
4   What do you think the 'mono' means in 'monomer'?
5   Name three plastics and give examples of how they are used.

6   Give two advantages and two disadvantages of plastics, compared with traditional materials.
7   Draw a diagram of an ethene molecule. Then show how ethene molecules join together when ethene is polymerised.

5.9 Shaping plastics

Washing-up bowls, toothbrush handles, shampoo bottles, milk crates, and plastic bags – they are all made of plastic, and they all have complicated shapes. How are these shapes made?

Two types of plastics

Most of the plastics that you are familiar with – polythene, PVC, polystyrene, nylon – have one thing in common. They go soft and runny whenever they are heated, and harden again when they cool down. They are called *thermoplastics*. ('Thermo' comes from a Greek work that means 'hot'.)

Other plastics like bakelite and melamine are heated once, when they are first made. Then they set hard and stay hard even if heated again. They are called *thermosetting plastics*.

Thermoplastics are ideal for making into complicated shapes, because they can be made soft and mouldable by heat. On the other hand, thermosetting plastics are best for things that have to withstand heat, like kitchen work surfaces, table mats, saucepan handles, and plastic cups and plates.

Extrusion

Sometimes, things made of plastic have to be very long, but the same shape all the way. Think of drain pipes, curtain rails, nylon thread, and the PVC covering for electric wires. These things are all made by a process called *extrusion*.

The plastic is heated up until it is soft and then squeezed through a shaped nozzle called a *die*, rather like the way you squeeze toothpaste out of a tube. The plastic comes out of the die the right shape, and is immediately cooled by cold air so that it goes hard and keeps the shape.

Figure 1 shows nylon being extruded in this way to make thin fibres for spinning and weaving into cloth.

Moulding

You couldn't make plastic bottles, dust bins, milk crates, or washing-up bowls by extrusion. Their shapes are too complicated. Instead you make them by *moulding*.

First, a mould of the required shape is made. Then it is filled with hot runny plastic, so that the plastic takes up the same shape. When the mould is cooled and opened, out comes hard plastic in the shape you wanted.

The hot runny plastic can be forced into the mould using a syringe or plunger, or blown in with compressed air, or sucked in by suction. Figure 2 shows a dustbin that has been made by forcing hot plastic into a mould. This method is called *injection moulding*. Figure 3 shows bottles that have been made by blowing plastic into a mould with compressed air. This method is called *blow moulding*.

Figure 1 Nylon fibres are made by extrusion.

Figure 2 Plastic dustbins are made by injection moulding.

Figure 3 These plastic bottles have been made by blow moulding.

Calendering

The sheeting for plastic bags is made by rolling hot polyethene between a series of giant rollers or calenders until it is the right thickness. The first set of rollers is heated so that the plastic is easily squashed. The final rollers are cooled so that the plastic goes hard again. The process is called *calendering*. Figure 4 shows a calendering machine.

Dip moulding

Some objects have to be plastic coated – like the handles of kitchen scissors and garden tools. For this they are dipped into a paste of plastic and then heated. Plastic boots ('wellies') are made by this method too. A mould in the shape of the boot is used. After cooling, the 'wellie' is peeled off the mould. Look at Figure 5.

Compression moulding

So far, you have seen ways of shaping thermoplastics. But electric light fittings and saucepan handles are made of thermosetting plastics. To make objects like these, the raw materials are put into a mould and then compressed and heated at the same time. Substances for kitchen work surfaces are made in a similar way. Layers of cloth and paper are soaked in liquid containing the thermosetting plastics, and then heated and compressed between hot metal plates. This is called *laminating*. Look at Figure 6.

Figure 4   A calendering machine is used to make plastic sheeting.

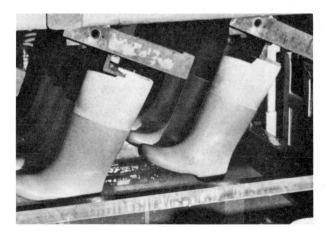

Figure 5   These boots were made by dip moulding.

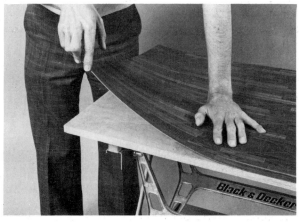

Figure 6   This man is putting a laminated surface onto the table.

Exercises

1   Name two thermoplastics and two thermosetting plastics.

2   How would you tell the difference between a thermoplastic and a thermosetting plastic in the laboratory?

3   What would be the best sort of plastic for making:
   a) a table mat?   b) a plastic doll?

4   What method would you use to make:
   a) a plastic table cloth?   b) the knob for a saucepan lid?   c) a plastic sandwich box?

Detergents

A product of sulphuric acid

In Unit 5.4 you read about sulphuric acid. Thousands of tonnes of this substance are used in the manufacture of detergents each year.

Detergents are *cleaning agents*. They are not just white powders or green liquids. Soap flakes and soap tablets also count as detergents. Detergents do two things. First of all they reduce the *surface tension* of water. Second, they dislodge particles of dirt and grease and carry them out of the material that is being washed.

You have all seen the effect of surface tension in water. It makes the water behave as though it had a skin on its surface. (Of course, there isn't really a skin there.) Try spreading a little water over a microscope slide. Does the water spread out? Or does it stay in blobs? Look at Figure 1. Surface tension stops the water 'wetting' the glass or spreading out over its surface. In the same way it stops water spreading out over your hands or dishes, if you wash them without soap.

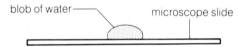

Figure 1   Surface tension keeps the water as a blob.

Now try adding a drop of detergent to the blobs of water. Figure 2 shows what happens to the blobs. The detergent reduces the surface tension and the water spreads out over the surface of the glass.

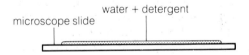

Figure 2   Detergent destroys the surface tension.

A specially-made molecule

A detergent is able to destroy surface tension because of the special structure of its molecules.

A detergent molecule is mainly a long chain of carbon atoms, which doesn't dissolve in water. This chain is called the water-hating or *hydrophobic* part of the molecule. But at one end of the chain, there is a group of atoms that do dissolve in water. This is called the water-loving or *hydrophilic* part of the molecule. Look at Figure 3. It shows a molecule of *sodium palmitate* which is a detergent made from palm oil.

You can see that the hydrophilic end has + and − signs on it, to show that the atoms or groups of atoms there are charged. In other words, they are *ions*. Ions dissolve in water. So when sodium palmitate is added to water, the hydrophilic end dissolves in the water and reduces its surface tension, allowing the water to thoroughly 'wet' the material that's being washed. The hydrophobic end of the molecule agitates the dirt particles and dislodges them from the material. It is also attracted to any grease or oil, and carries it out of the material. Look at Figure 4.

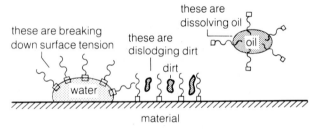

Figure 4   The washing action of a detergent

Two types of detergents

There are two sorts of detergents. First there are the *soapy* detergents, like the ones you use for washing and bathing your skin. We just call them *soaps*. They are made from fats and oils like beef and mutton tallow, and palm, olive, coconut, and peanut oils. The second sort are *soapless detergents*. They are used as washing-up liquids and shampoos, and for washing clothes. They are made

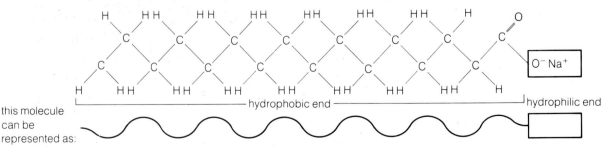

Figure 3   A molecule of the detergent sodium palmitate

from sulphuric acid and chemicals obtained from crude oil. In both sorts of detergents the molecules have the same type of structure – a hydrophobic chain with a hydrophilic group of atoms at one end.

The big disadvantage of soapy detergents is that their washing action is stopped by *hardness* in water. No tap water is pure, since water is such a good solvent. It dissolves all sorts of compounds from rocks and soil, on its way to the water works. If it flows over limestone rocks, or through soil containing magnesium compounds, then it will contain small amounts of calcium hydrogencarbonate, magnesium sulphate, and calcium sulphate. These compounds destroy the soap by reacting with it and converting it into insoluble compounds that float to the surface. We call the insoluble compounds *scum*.

Water that contains these calcium and magnesium compounds is said to be *hard*. In hard water areas, a lot of soap is needed to get a good lather, and a lot of scum is formed as well.

Soapless detergents do not react with calcium and magnesium compounds in this way, so their washing performance is not affected by hard water.

The manufacture of soapless detergents

The first man-made detergent was made in France in 1831. In this country, the first one was Persil, made in 1901. By the 1950s the detergent industry was well under way, using the chemicals extracted from crude oil.

The type of chemical that is used is called an *alkylbenzene*. This is reacted with *oleum*, which is super-concentrated sulphuric acid, and the alkylbenzene becomes *sulphonated*. Next, water is added to dilute it. Then sodium hydroxide is added to neutralise the acid. The result is a detergent.

A detergent is mixed with several other chemicals, before it comes out of the reaction vessel. *Sodium tripolyphosphate* is added as a *builder*. It builds up or increases the power of the detergent because it can loosen dirt particles. Another builder called *sodium carboxymethyl cellulose* is used to make sure that the dirt particles stay suspended in the water and do not go back onto the material. *Sodium perborate* is added as a *bleach*. *Sodium silicate* is used as a *filler* to make sure that the powder stays dry and runs smoothly. Persil contains these last two compounds. Can you see how it got its name?

This mixture of chemicals is dried with hot air. It is then powdered and put into packets.

Liquid detergents are made using slightly different chemicals, but the idea is the same.

Whiter than white?

New white clothes look really white because the material can reflect the blue light in sunlight. But after a few washes they begin to go a bit yellow. This problem is solved by compounds called *fluorescers* that are added to some detergents. Fluorescers absorb the invisible *ultraviolet light* from sunlight, and give it out again as slightly blue light. During a wash, the fluorescers are attracted to the clothes, and get absorbed by them. So in daylight, the white clothes get back that new-white look again. (At least in theory!) Look at Figure 5.

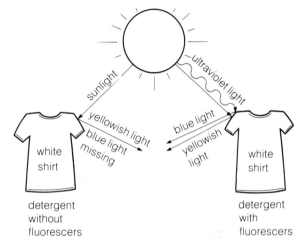

Figure 5 Fluorescers in a detergent absorb ultra-violet light and give it out again as slightly blue light. This makes white clothes look whiter.

Molecules that munch?

Enzyme detergents contain compounds that react with substances such as blood, sweat, egg-stains, and milk. They break them down into water-soluble compounds so that they can be washed out easily. Non-enzyme detergents have to rely on a bleach to whiten the stain. Figure 6 shows an enzyme ('biological') detergent.

Figure 6 Some detergents contain enzymes to break down food and body stains.

Topic 5 Exercises

1 What name is given to the rock from which a metal compound is obtained?

2 Name one native metal.

3 What is the purpose of the oil–water mixture, in the extraction of copper?

4 Write the word equation for the roasting of copper sulphide in air.

5 What method is used to make crude copper pure?

6 Name three important uses for copper.

7 What three substances are put into a blast furnace to make iron?

8 How is the blast furnace heated up?

9 Write the equation for a reaction of carbon with oxygen.

10 Haematite has to be 'reduced' to give iron. What does that mean?

11 Write the word equation for the reduction of haematite with carbon monoxide.

12 What is slag?

13 Why is cast iron not used for making bridges or girders?

14 How is iron turned into steel?

15 Name one alloy steel. What is it used for?

16 Name a metal that is extracted by electrolysis.

17 What is limestone used for in Britain?

18 How is quicklime made?

19 What would you see when water is added to quicklime? Write the equation for the reaction and name the compound formed.

20 What is lime water? How is it used to test for carbon dioxide?

21 Outline the way in which cement is made.

22 How much sulphuric acid is made in Britain each year, and what is it all used for?

23 Describe the Contact process for making sulphuric acid. Mention the starting materials, the conditions, and the way in which the sulphuric acid is extracted at the end.

24 Describe what you think you would see if you added some drops of concentrated sulphuric acid to glucose, $C_6H_{12}O_6$.

25 What precautions must you take when you are diluting concentrated sulphuric acid?

26 How would you expect dilute sulphuric acid to react with: a) magnesium?
b) copper? c) sodium carbonate?
To answer these questions, you may have to revise some work you did earlier.

27 Plants need nitrogen to grow. Where do they get the nitrogen from?

28 In the Haber process, what steps are taken in order to get as much ammonia as possible?

29 a) Some ammonia solution is put in a test tube and two drops of litmus solution are added. What colour will the litmus turn?
b) Next, some dilute hydrochloric acid is dripped into the test tube. What will happen to the litmus? What sort of reaction will take place and what compound will form?

30 What would you expect to happen when dilute nitric acid is added to: a) copper oxide?
b) magnesium carbonate? c) copper?

31 What is meant by the word 'synthetic'?

32 Give one example of a monomer.

33 What does 'poly' in the word 'polymer' mean?

34 Why is polystyrene a good insulator?

35 Name the plastic that is used to make:
a) glue b) string c) records d) safety glass.

36 What is meant by the term 'extrusion'?

37 What is the difference between a thermoplastic and a thermosetting plastic?

38 Which type of plastic would you use to make:
a) a table mat? b) the handle on a saucepan?
c) plastic chessmen?

39 Give three advantages and three disadvantages of plastics over traditional materials.

40 Here are some observations about a compound and its reactions. Read them carefully and see if you can work out what the compound is. You will need to use information you learned in other Topics.

i) The compound is a green powder.
ii) It fizzes when added to dilute sulphuric acid and a gas is given off which turns lime water cloudy.
iii) When the compound is heated, it turns black and gives off the same gas as in (ii).
iv) The black powder left after heating is mixed with carbon and heated again. A reduction reaction takes place. The product of this reaction is a pinkish metal.
v) The metal has no reaction with dilute hydrochloric acid, but it dissolves in nitric acid to form a blue solution and brown fumes of nitrogen oxide.

What is this green powder?

41 Draw the apparatus that you would use for the electrolysis of molten potassium iodide. Label all the parts and say which electrode is the cathode and which is the anode. Describe what will happen when the electricity is switched on.

42 Say why it would be unwise to make a saucepan handle out of polythene and a curtain rail out of bakelite.

Oil is an important source of energy. Most of Britain's oil comes from under the North Sea.

6.1 Coal

In the year 1981–82, over 124 million tonnes of coal were mined in Britain. Did it all go up in smoke?

What is a fuel?

As far as a chemist is concerned, a *fuel* is a chemical which reacts with oxygen when it is heated, giving out energy. The energy may be in the form of heat, which can boil water; or light, which can light up a room; or hot, expanding gases, which can drive a piston up and down in a car engine. In each case, the fuel is involved in a *combustion reaction*. In other words, the fuel *burns*.

How much energy do you use?

Scientists estimate that in 1977 each person in Britain used energy equivalent to the amount obtained from burning 6 tonnes of coal. By the year 2000, you may each be using twice as much.

You use energy in the form of *heat* from burning oil, coal, and gas; *light* from electricity (which is also made from coal, oil, or gas); *movement* (transport uses petrol and diesel); and locked up in thousands of consumer products like *plastics* and *foodstuffs* that used fuels in their production.

A lot of energy is wasted by letting heat escape or by throwing away goods that could be recycled. This means that the fuels used in the first place were wasted.

The rest of this section deals with one important fuel – *coal*.

What is coal?

Millions of years ago, our coal was alive and growing, in the form of trees and plants. When these died, their bodies sank into the ground, and were crushed when the Earth changed its shape during earthquakes and volcanic eruptions. Gradually, the dead plants and animals were compressed and turned into coal.

Coal is now found in layers or seams, often deep in the ground, in parts of the country that were covered by forests millions of years ago. Figure 1 shows the main coal fields in Britain. Because coal was once a living thing, it contains the elements carbon, hydrogen, and oxygen, as do all living things.

Coal mining

Some coal seams are very near the surface, so that the coal can be dug out with mechanical diggers. This method of mining is called *open cast* mining. However, most seams in Britain are deep in the ground and *mines* have to be constructed. In modern mines, tunnels or roads are cut into the rock. Then mechanical cutters are sent into the tunnels, to slice into the coal seam. The broken coal is loaded onto conveyors which carry it to the surface. The roofs of the tunnels are held up by special supports which can be made to 'walk' by the miners who operate them. Figure 2 shows a *coal shearer* which shears coal from the seam.

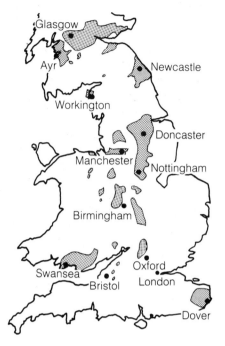

Figure 1 This map shows the coal fields of Britain.

Figure 2 This mechanical cutter is called a coal shearer. It shears coal from the seam.

The uses of coal

In the year 1981–82, the National Coal Board estimates that over 124 million tonnes of coal were mined, at a cost then of about £36.60 a tonne. Figure 3 shows the main uses of that coal. Most of it was burned in power stations. The heat energy from the burning coal was used to make hot, expanding steam. The movement energy in the steam was transferred to movement energy first in a turbine, and then in a generator. Finally, it was turned into electrical energy for use by everyone in the country. You can read more about the generation of electricity in Unit 6.4.

Some of the coal was used to make coke. For this, the coal is first heated in large ovens called *retorts*, without letting it burn. Gas called *coal gas* is driven off, together with ammonia solution, tar, and other chemicals. Coke is left behind. Coke is needed to make iron and steel, and for smokeless fuels. Coal gas used to be a common fuel for public use, but North Sea gas is now used because it is hotter and cleaner. The other substances – including ammonia and tar – are turned into chemicals for all sorts of products.

Figure 4 shows how coal is split up and what the different parts are used for.

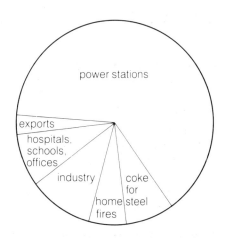

Figure 3 This pie chart shows where all the coal mined in Britain went to in 1981–2.

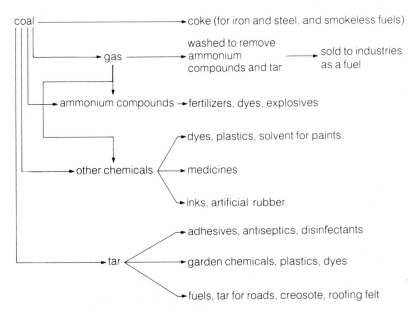

Figure 4 Coal is split up into coke, coal gas, ammonium compounds, other chemicals, and tar. Some of the uses of these substances are shown in this diagram.

About three-quarters of the coal produced in Britain is transported by rail. This coal is being delivered to Fiddler's Ferry power station in Cheshire.

Exercises

1 What element does a fuel need in order to burn?

2 Can you suggest why coal is called a 'fossil fuel'?

3 When the National Coal Board estimates the cost of mining coal, what sorts of things do you think it takes into account?

4 a) Name four important substances that are made when coal is heated.
 b) Name one use or by-product of each of these substances.

5 What gases will form when you burn a fuel that contains hydrogen, carbon, and oxygen?

105

6.2 Oil

Like coal, oil is called a fossil fuel. It was made millions of years ago from the bodies of tiny sea animals.

Searching for oil

Geologists know that certain types of rock, in certain parts of the world, may contain oil – possibly!

Millions of years ago, when most of the life on Earth was in the sea, the bodies of the dead sea animals piled up on the sea bed. There they were gradually buried and compressed by layers of mud and silt, and slowly turned into oil. Now these layers are deep in the earth, with the oil trapped inside them. Over the centuries they have been buckled and folded by the changes that have taken place in the Earth's structure. Look at Figure 1.

By studying the contours of the ground from an aeroplane, a geologist can spot areas with the sort of structure that suggests oil. Next, the geologist will try to identify the layers under the ground in these areas, by a process called a *seismic survey*. Look at Figure 2. The geologist fires a series of small explosions in the ground and measures the shock waves as they rebound from the layers of rock. If he gets encouraging results, he will then drill down and extract samples of the rock. He will examine these for fossils of the animals that made oil, and analyse the rocks to see if they are oil-bearing. All this information will help him to decide the most important thing of all – whether or not to drill a well.

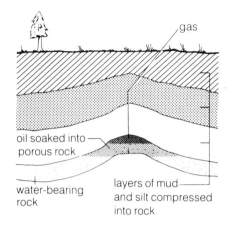

Figure 1 Oil might be trapped like this under the ground.

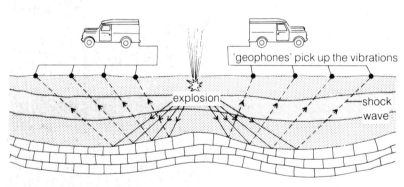

Figure 2 A geologist uses a seismic survey to try to identify the layers under the ground.

Drilling for oil

Before drilling begins, an oil company will already have spent a lot of money surveying the ground. It must spend millions of pounds more sinking a well, with the risk that there may be no oil present at all. Figure 3 shows the tower or *oil derrick* that is set up, so that the *drill bit* can be driven into the ground. The tower and pulley are necessary because the workers have to connect 30-metre lengths of pipe to the drill, stage by stage, as it sinks down into the earth. Figure 4 shows workers making a connection.

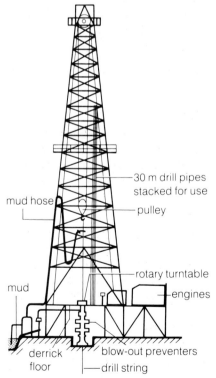

Figure 3 An oil derrick

The drill and piping together are called the *drill string* and may weigh as much as 100 tonnes. The drill has to be changed from time to time, either because it has worn out or because the type of rock has changed. See Figure 5. This means that the drill string has to be pulled up and disconnected, pipe by pipe. The job may take many hours to complete.

If oil is found, the well must be capped with valves before the oil can be pumped out. If gas is found too, the pressure can be dangerously high, and special valves called *blow-out preventers* must be used.

Most of Britain's oil comes from under the North Sea. So the problems of surveying and drilling and pumping are even greater, since so much has to be done underwater, and often in very cold and stormy conditions. The oil has to be piped ashore to *oil terminals*, or filled into tankers at sea. Look at Figure 6.

Figure 4 These men are joining two drill pipes which make up part of the drill string.

Figure 6 An oil tanker at an oil terminal

Gas

In addition to the oil under the North Sea, *natural gas* is found there too. Natural gas is mainly *methane*, and it was formed in much the same way as oil. In fact, methane is made whenever vegetation rots or gets broken down. This happens in marshes, at sewage works, in coal mines (where it is explosive and dangerous), and even inside you!

The gas is detected and drilled for in the same way as oil, and is piped ashore to *gas terminals*. From these, it is distributed throughout the country as a fuel. Methane is a good fuel because it burns with a hot flame and because, in a good supply of air, it burns cleanly with no harmful products.

Figure 5 A drill bit has to drill through layers of rock, sometimes thousands of metres thick, in the search for oil. Sometimes a new bit is needed every day. These drill bits are worn-out.

Exercises

1 Why is oil called a fossil fuel?

2 What is a seismic survey?

3 Why does an oil derrick need to have a pulley and a powerful engine?

4 Try to explain why methane is dangerous in coal mines.

5 Explain why drilling for oil is such a long and expensive business.

6.3 Refining oil

Crude oil is a mixture of hundreds of different compounds. These must be separated in an oil refinery.

Oil, a mixture

There are hundreds of different compounds in crude oil. They are all *organic* compounds. That means they were all made from living things, and their molecules all contain carbon atoms. The simplest of these compounds are called *hydrocarbons* because they contain only hydrogen and carbon atoms.

You saw earlier that *methane* occurs as natural gas. It is also found dissolved in oil. It is the simplest hydrocarbon of all. Figure 1 shows a methane molecule. It contains just one carbon atom, joined to four hydrogen atoms. It is small and light, which explains why methane is a gas. Other gases similar to methane are also found dissolved in oil. Their names are *ethane*, *propane*, and *butane*. The structure of their molecules is shown in Figure 2. Under pressure, these gases turn into liquids. This makes them very useful as fuels, for liquids are much easier to package and carry around than gases are. Figure 3 shows liquefied propane for use as fuel.

Crude oil also contains dissolved solids, but most of the compounds in it are liquids. Some of these have very high boiling points. This is because their molecules are large and heavy, with as many as 50 or 60 carbon atoms in them. *In general, the more carbon atoms there are in a molecule, the higher the boiling point of the compound will be*. The number of carbon atoms also determines the nature of the liquid. If the number is small, the liquid tends to be runny and *volatile* – it evaporates easily (like petrol does). If the number is large, the liquid tends to be thick and oily.

Separating the mixture

The compounds in oil are separated into groups, depending on their boiling points (and hence on the sizes of their molecules). This is called *refining* the oil.

For example, one group contains the compounds that boil between 220 °C and 250 °C. This group is used as *diesel oil*. Compounds in it have from 15 to 20 carbon atoms in their molecules.

The table below shows the groups of compounds that are separated from oil, and the sorts of temperatures at which they boil.

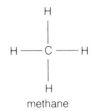

Figure 1 Methane is the simplest hydrocarbon. It contains one carbon atom and four hydrogen atoms.

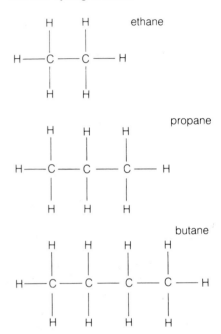

Figure 2 Ethane, propane, and butane are gases dissolved in oil.

Figure 3 Liquefied propane is a useful fuel. 'LPG' stands for 'liquid petroleum gas'.

name of group	boiling points of liquids	number of carbon atoms in molecule
refinery gas	less than 40 °C	1–4
petrol and naphtha	40–170 °C	4–12
kerosene	150–240 °C	9–16
diesel oil	220–250 °C	15–25
lubrication oils	250–350 °C	20–70
bitumen	above 350 °C	60 and above

How refining is carried out

A method called *fractional distillation* is used to refine the oil. You read about this method in Unit 1.6. It works because each group of compounds has a different range of boiling points.

The sort of apparatus that is used in the laboratory is not suitable for separating thousands of litres of oil. Instead a *fractionating tower* is used. (The different groups of compounds are called *fractions*.) Figure 4 is a photograph of a fractionating tower and Figure 5 shows what goes on inside it.

Figure 4 A fractionating column (also called a fractionating tower)

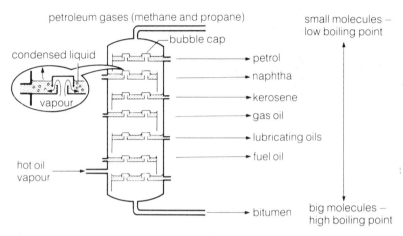

Figure 5 Crude oil is separated into fractions in a fractionating column.

First the oil is heated. Methane and the other dissolved gases bubble off very quickly. Then the liquid boils, giving a mixture of vapours. The vapours flow up the tower and pass through *bubble caps*. In these, the higher boiling-point vapours condense to liquids, but the others carry on as gases. This is repeated at intervals up the tower, until the oil is separated into five or six different fractions.

The fractions are removed from the tower. They may be fractionally distilled again if necessary, to split them up into even smaller groups of compounds. Some, like naphtha, are mixed with steam and passed over hot catalysts so that their large molecules get broken down into smaller ones. This is called *catalytic cracking*. Most of our petrol is made by catalytic cracking.

Where does it all go?

Figure 6 shows how the fractions from oil are used. More than 80% of the oil ends up as fuel, in the form of fuel oil for power stations, central-heating oil for homes and offices, petrol for cars, diesel oil for lorries, and paraffin for jets. The rest is used to make chemicals.

fraction from crude oil	use
bitumen	road surfaces waterproof materials, e.g. damp coursing, asphalt roofing
fuel oil	fuel for ships and power stations
lubricating oils	for machinery waxes and polishes
gas oil	diesel fuel for lorries
kerosene	paraffin, jet fuel central heating oil
naphtha	chemicals: nylon PVC polystyrene fertilizers insecticides dyes
petrol	fuel for cars
petroleum gases	bottled gas eg. Calorgas

Figure 6 The products of crude oil refining have many important uses.

Exercises

1 What is a hydrocarbon?

2 The gas butane is used as camping gas. But if you shake a can of camping gas, you can hear a liquid inside. Can you explain why?

3 How does the number of carbon atoms in its molecules affect the boiling point of an organic compound?

4 What name is given to the process of separating a mixture of liquids by boiling?

5 Draw a bubble cap, and explain what happens when oil vapours pass through it.

6 List the different fractions from oil, and say how each is used.

6.4 Nuclear energy

Most of the time in chemistry, it is useful to think of atoms as small spheres. But for nuclear scientists, it's an inside job. . . .

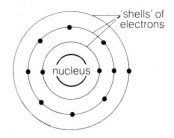

Figure 1 Inside an atom, electrons move around a central nucleus.

Inside the atom

An atom is made of a central *nucleus*, in which most of the mass of the atom is contained, and small packets of energy called *electrons*. The electrons exist in layers or *shells* round the nucleus. Look at Figure 1. The more energy an electron has, the further its shell is from the nucleus. The shells are not really there. They are just the paths taken by the electrons as they move around.

The nucleus contains two types of particles. These are *protons* and *neutrons*. Each is nearly 2000 times heavier than an electron.

Nuclear fission

In some big atoms, the nucleus is *unstable*. This means it will easily split into two smaller nuclei. (*Nuclei* is the plural of 'nucleus'.) This may happen spontaneously, without any help from outside the atom. It will also happen if another particle, such as a neutron, is made to collide with the atom.

The scientific word for 'splitting' is *fission*, so when the nucleus of an atom splits, the event is called *nuclear fission*. When nuclear fission takes place, a great deal of heat is given out.

Nuclear fission was first made to happen with *uranium* atoms. When a neutron hits a uranium atom, the atom splits into two smaller atoms, and ejects three fast-moving neutrons. The three neutrons may immediately collide with three more uranium atoms and split them. Each of these will in turn eject three more neutrons, which may collide with three more uranium atoms, and then these . . . but look at Figure 2.

This fast-growing reaction is called a *chain reaction*. If the piece of uranium is big enough, the heat from the chain reaction quickly gets out of control. A vast ball of fire and hot expanding gas is formed, and it spreads in every direction. In other words, an atom bomb has gone off! Besides the heat, the chain reaction also releases energy rays and dangerous bits of atoms, which we call *radiation*.

Some of this radiation dies away quickly but much of it gets attached to the dust and debris that are sucked up into the mushroom cloud when the bomb goes off. This radioactive dust, now called fallout, drifts over hundreds of miles and slowly falls to earth, carrying its radiation with it. The first atom bomb was dropped on Hiroshima, in Japan, in 1945. It killed 80 000 people within a day, and another 60 000 died within a year from the effects of the radiation.

The same chain reaction occurs in a nuclear power station, but there it is kept very carefully under control. The heat from it is used to generate electricity. The container in which the reaction actually takes place is called a *nuclear reactor*.

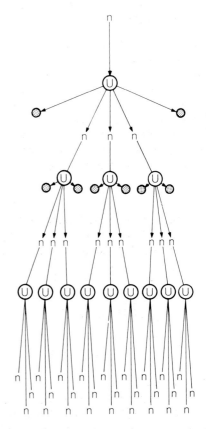

Figure 2 When a neutron splits a uranium atom, a chain reaction begins.

Modern reactors

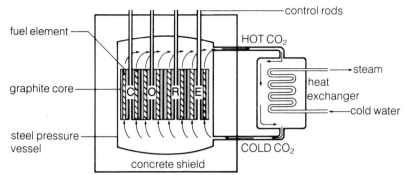

control rods
fuel element
HOT CO₂
graphite core
CORE
steam
heat exchanger
cold water
steel pressure vessel
COLD CO₂
concrete shield

Figure 3 Inside a Magnox nuclear reactor, carbon dioxide circulates to remove heat from the core.

Figure 3 shows the inside of a Magnox reactor. There are nine of these dotted around Britain, producing electricity for the National Grid. The *core* of the reactor is made of *graphite*, with thousands of holes drilled through it. Down into the holes are lowered the *fuel elements*. These are narrow tubes, each about a metre long and filled with specially prepared uranium. Look at Figure 4. The tubes themselves are made of a magnesium alloy – which is why the reactor is called a *Magnox* reactor. The graphite acts as a *moderator*. When neutrons start to collide with the uranium atoms, the graphite slows them down so that they do the job more efficiently. To stop the chain reaction getting out of hand, *control rods* made of *boron steel* can be lowered into the core. These will absorb neutrons and thus make the reaction slow down.

The core is enclosed in a steel pressure vessel. This is in turn surrounded by a thick casing of reinforced concrete, to keep the dangerous radiation in. Because the core gets very hot, cold carbon dioxide gas is pumped round it to carry the heat away. The hot gas is then made to flow over pipes of water in a *heat exchanger*. Here it cools down again, by passing its heat to the water, which changes into steam. The cool carbon dioxide is led back to the core to fetch more heat. It cannot be allowed to escape, because it is *radioactive* – that is, it carries radiation.

Figure 4 Fuel elements for Magnox reactors are made from uranium metal rods enclosed in magnesium alloy cans.

The steam is carried away from the heat exchanger and made to turn a *turbine*. This is just like a windmill. The high-pressure steam hits the blades of the turbine and makes it spin. The turbine is connected to a coil of wire in a magnetic field. So the coil spins too, and the result is electricity. Look at Figure 5.

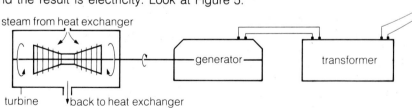

steam from heat exchanger
generator
transformer
turbine
back to heat exchanger
National Grid electricity supply

Figure 5 From steam to electricity

Exercises

1 Name the three particles found in an atom and draw a diagram to show how they are arranged.

2 Explain what is meant by the term 'nuclear fission'.

3 What happens in a chain reaction?

4 Explain why a nuclear reactor does not explode like a bomb.

5 What job does the carbon dioxide do, in a Magnox reactor?

Brazil's answer to the oil problem

Many problems – but some advantages

Brazil is a huge country of over 3 million square miles. It is also quite a poor one. During this century, it has borrowed billions of pounds from richer countries to help improve its industry and farming. The interest alone on this borrowing is about 4 billion pounds a year. A lot of money!

Brazil does have oil wells of its own, on shore and in coastal waters. But these produce only 18% of the oil it needs. A modern industrial nation needs lots of oil for transport and power stations and the manufacture of fertilizers and other chemicals. However, Brazil has three things in its favour: lots of space for farming; many natural forests for raw materials; and a warm climate. One crop that needs both space and heat is *sugar cane*. Look at Figure 1.

Figure 1   Fully-grown sugar cane

The Proalcool project

As early as 1931, Brazil was attempting to produce *ethanol* (alcohol) from sugar cane. During the Second World War, it was forced to use ethanol as a fuel because its oil imports were completely stopped.

In 1975, the *Proalcool project* started. Its aim was to blend ethanol with petrol (made from crude oil) to give a good fuel, so that Brazil could cut down on oil imports. By 1978, over 1.6 million litres of petrol had been saved in this way and today, most of the petrol pumps in Brazil sell petrol mixed with 20% ethanol. Look at Figure 2. This was chosen as the best mixture because it did not require any changes to the petrol pumps or car engines.

The Proalcool project did not stop there. By 1985,

Figure 2   Most petrol pumps in Brazil sell petrol mixed with ethanol (alcohol). 'Movido a alcool' on the car means 'runs on alcohol'.

it aimed to have all cars and lorries running on pure ethanol. This would need 11 billion litres of ethanol a year. There are already more than 200 pumps serving pure ethanol in Brazil, and some 300 000 cars are already using only this fuel. The cost of converting your car from petrol to ethanol is about £200, and Brazilian factories are already producing new cars that run on ethanol.

Apart from the reduction in cost, there are other big advantages. Ethanol produces just as much energy as petrol. Moreover, it does not need the many additives that petrol must have, such as lead compounds. And when it is burnt, its exhaust fumes are much cleaner.

Now for some economics

At first glance, the solution seemed simple. Brazil would convert some of its sugar into ethanol. It would import less oil and so save money. Its own industries would be improved and developed and motorists would get fuel at about half the cost of petrol.

However, when studied in more depth, things were not quite so simple. The first ethanol came from distilleries built next to sugar refineries. They turned spare sugar into ethanol at the rate of about 5.5 billion litres a year. If this figure was to double by 1985, some of the sugar intended for export would have to be used for ethanol instead. When the project first started, the amount of Brazilian sugar on the world market decreased. The resulting shortage caused the price of sugar to shoot up. Brazil then found it more profitable to sell sugar and buy oil! Quite the reverse of what had been intended. Brazil would obviously have to grow more sugar so that both its needs could be satisfied. In fact it would have to double the size of its sugar cane crop.

One group of people who were not too happy with this solution was the industrialists. They pointed out that much of the imported oil was used to make chemicals, and plastics like polythene and PVC. These would have to be made from ethanol instead, if oil imports were cut. And that meant even more ethanol would have to be produced. Besides, crude oil is needed for the diesel fuel used by lorries and buses. Diesel fuel can be mixed with only 7% ethanol without needing changes to engines. However, scientists are already working on this problem. They are trying oils made from peanuts, soya beans, and sunflower seeds! Up to 33% of these oils can be mixed with diesel.

The biology and chemistry of it all

Sugar cane takes 1½ years to grow to maturity, and once grown it can only be cut twice in its lifetime. The canes are crushed in big mechanical mangles and their syrup squeezed out. The sugar is then extracted from this. If ethanol is required, the syrup is diluted with water and put into vats with yeast. Then the fermentation process starts. Sugars with complicated structures are broken down into the simplest sugar, *glucose*. Substances called *enzymes* in the yeast act on the glucose, and ethanol and carbon dioxide are produced:

$$\underset{\text{glucose}}{C_6H_{12}O_6} \xrightarrow{\text{yeast}} \underset{\text{ethanol}}{2C_2H_5OH} + \underset{\text{carbon dioxide}}{2CO_2}$$

When the amount of ethanol has reached about 9% of the mixture, the yeast dies and fermentation stops. The next stage is to remove the 91% of water in the mixture. Figure 3 shows the sort of apparatus that would be used in the laboratory to do this. It is the apparatus for *fractional distillation* that you met already in Unit 1.6. The fermented mixture of ethanol and water is put in the flask and heated. A mixture of ethanol vapour and

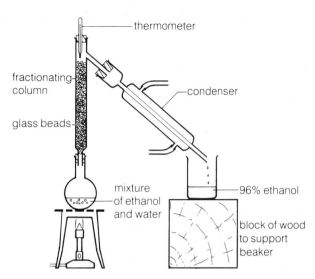

Figure 3 This is the apparatus used for the laboratory distillation of ethanol. The apparatus used in distilleries works in a similar way.

steam rises up the fractionating column. The vapours condense on the glass beads in the column, but the ethanol vapour is able to get higher up than the steam. (This is because ethanol boils more easily. Its boiling point is 78 °C compared with 100 °C for steam.) Eventually, only ethanol vapour reaches the top. It goes into the condenser where it turns back to liquid ethanol. This method produces a mixture of 96% ethanol and 4% of water. To make the ethanol absolutely pure and water-free, it is mixed with a liquid called *benzene* and distilled again.

The apparatus used in distilleries is of course much larger than the laboratory apparatus, but it works in much the same way. The husks from the sugar cane are burned to provide the heat for the distillation.

What about other crops?

Brazil is busy trying other crops to see if they will produce ethanol too. *Cassava roots* are easily grown in poor soil and several project distillation plants have been set up at farms that grow this crop. Some of them have produced up to 1000 litres of ethanol a day. The fermentation of cassava is a little more difficult than that of sugar, because cassava has a more starchy structure. But it has other advantages. The peel can be used as cattle food. The residue from the fermentation can also be fed to animals, or rotted down to make methane (another fuel), or even used to make lubricating oils. The carbon dioxide is not wasted either. It is used to make dry ice for refrigeration, or put into fizzy drinks.

Topic 6 Exercises

1 What are fuels, and what are they used for?

2 By following the steps outlined below, complete this equation for the combustion of propane, C_3H_8:

propane + oxygen → carbon dioxide + steam
C_3H_8 + O_2 →

a) First, look at the formula of propane. It tells you how many molecules of carbon dioxide there should be on the right-hand side of the equation. Write them in.

b) The formula of propane tells you there are 8 atoms of hydrogen in the reaction. How many molecules of water can come from 8 atoms of hydrogen? Write them in.

c) How many molecules of oxygen are needed, to balance the equation? Write this number in on the left-hand side.

d) Check that the equation is now balanced.

3 Give examples to show how energy from a fuel can be converted into: a) movement energy b) sound energy c) electrical energy.

4 What is meant by the term 'combustion'?

5 What is coal? How was it formed?

6 Outline two ways in which coal is mined.

7 How much coal was mined in 1981–82? Work out how much the whole lot cost.

8 What is coke used for in the iron and steel industry? (You will need to turn back to Unit 5.2.)

9 Make a list of ten objects in your home. How many of them are made from chemicals that came from coal?

10 How do geologists set about finding oil? Outline the steps they take.

11 What is a drill string? Why is it such a long and difficult job to move the drill string in and out of the ground?

12 What is a blow-out preventer and why may it be needed on an oil derrick?

13 Why is methane such a good fuel?

14 What is naphtha used for?

15 How is the boiling point of a liquid related to the number of atoms in its molecules?

16 What happens in catalytic cracking?

17 From which fraction of crude oil are waxes and polishes made?

18 What is the centre of an atom called?

19 What are electrons?

20 How much heavier is a proton than an electron?

21 What are the shells in an atom?

22 What is meant by the term 'nuclear fission'?

23 Name an element that undergoes nuclear fission.

24 Draw a diagram to show how a chain reaction develops in an element undergoing nuclear fission.

25 How does a nuclear explosion happen?

26 What is the difference between a nuclear explosion and the nuclear reaction in a power station?

27 Draw a diagram of a Magnox reactor and label the parts, explaining the job of each part.

28 Why is a nuclear reactor surrounded by thick concrete?

29 In a Magnox reactor, carbon dioxide gas is used as the coolant. Why must the gas not be allowed to escape into the air?

30 Explain how the energy from nuclear fission in a nuclear reactor is used to make electricity.

31 The graph below shows how much energy is predicted to be available and how much will be demanded in the near future. Study the information it gives and answer the questions that follow:

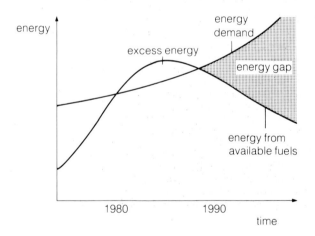

a) Why do you think Britain had more energy than it needed in 1980, although it hadn't enough in the 70s?

b) Why is there likely to be an energy gap after 1990? What will have run out by then?

c) Why does our energy demand keep on increasing?

Acknowledgements

Anglia Water Authority: p. 27;
Aspreys & Co. plc: pp. 71 (top), 78 (bottom);
Associated Press: p. 24 (top);
BBC Hulton Picture Library: pp. 50, 89 (bottom);
B.I.C.C: p. 71 (bottom right);
Biofotos: pp. 41, 43, 58 (centre), 64 (top);
B.P: pp. 9, 107 (top right);
British Gas: p. 55;
British Museum: p. 78 (top and centre);
British Oxygen: pp. 24 (2nd from top), 71 (bottom right);
British Steel Corporation: p. 51;
British Sugar plc: p. 8;
Maurice Broomfield: p. 103;
Camera Press: pp. 16 (left), 26 (bottom), 107 (bottom right);
Camera Talks: p. 30;
Costains: p. 84;
De Beers: pp. 22, 45 (top right), 80;
Dunlop: p. 99 (top);
Esso: pp. 77, 107 (left), 109;
Explosives and Chemical Products Ltd: pp. 83 (top), 90 (2nd from top);
Fiat Auto UK Ltd: p. 112 (right);
Formica Ltd: p. 99 (bottom left);
Geoscience Features: pp. 10 (left), 19, 44 (2nd from left);
Glass Manufacturers Federation: p. 17;
The Guardian: pp. 36, 54;
Philip Harris: pp. 4, 5, 11;
ICI: pp. 87 (bottom), 90 (2nd from bottom), 93;
ICI Agriculture Division: p. 63 (left);
Jamaica Tourist Board: p. 26 (top);
JCB: p. 37 (top);
Keystone Press: p. 24 (2nd from bottom), 57, 72 (left), 71 (centre);
Mansell Collection: pp. 34 (top), 79 (top);
Monsanto: p. 87;
Mullards: p. 79 (bottom);
N.A.S.A: pp. 23, 38 (2nd from right);
National Trust: p. 85;
N.C.B: pp. 104, 105;
Permutit-Boby: p. 21 (bottom);
R.K. Pilsbury: p. 89 (top);
The Science Museum, London: p. 79 (centre);
Tate and Lyle: p. 112 (left);
Tate Gallery: Rodin The Kiss, p. 22;
U.K.A.E.A: pp. 34 (bottom), 111;
U.S.I.S: p. 35 (left);
C. Ward-Perkins: p. 10 (centre right);
Jerry Wooldridge: p. 21 (centre);
West Midlands Fire Service: p. 25;
Whitechapel Bell Foundry: pp. 74, 75;
Zambia Information Services: p. 38 (2nd from left).

Glossary

These are a few of the important words that you will meet when you start chemistry. You should be able to explain what they mean and give examples of each of them.

Atom
Atoms are tiny particles of which everything is made. Each element (see below) has its own atom. All the atoms in an element are the same, but they are different from the atoms of other elements. In ordinary chemical reactions, atoms cannot be split into anything smaller.

Molecule
A molecule is a group of two or more atoms joined together. For example, a molecule of oxygen consists of two oxygen atoms, and a molecule of carbon dioxide consists of one carbon atom and two oxygen atoms.

Element
An element is a pure substance made up of only one sort of atom. Carbon, sulphur, iron, and aluminium are examples of elements.

Compound
A compound is a pure substance made up of two or more elements joined together. Examples are copper oxide (made from copper and oxygen) and sulphuric acid (made from hydrogen, sulphur, and oxygen).

Symbol
Each element has its own symbol which is made of a letter or letters from its name or from some other word. For example, carbon has the symbol C but iron has the symbol Fe from its Latin name.

Periodic Table
This is a list of all the elements represented by their symbols. The order in which they appear depends partly on how heavy their atoms are (hydrogen has the lightest) but also upon the structure of the atoms.

Formula
A formula is a compound's symbol. It shows what elements the compound contains and the numbers of each sort of atom. For example, the formula of sulphuric acid is H_2SO_4. This formula tells you that every molecule of sulphuric acid contains two hydrogen atoms, one sulphur atom, and four oxygen atoms.

Equation
An equation is a chemical reaction written like a sentence, using symbols and formulae. For example:
   'iron reacts with sulphur to form iron sulphide'
can be written as
   $Fe + S \rightarrow FeS$

Ion
An ion is an atom or molecule that has been given a positive or negative charge. For example, an oxygen atom with two negative charges on it is called an oxide ion. A hydrogen atom with one positive charge on it is called a hydrogen ion.

Acid

An acid is a compound which dissolves in water to produce a solution containing hydrogen ions. Acids turn litmus paper red. Sulphuric acid and hydrochloric acid are generally used in the laboratory.

Base

A base is a compound which is an oxide or a hydroxide of a metal. Bases neutralise acids. Copper oxide and aluminium hydroxide are examples of bases.

Alkali

Bases that dissolve in water are called alkalis. Examples are sodium hydroxide solution and ammonia solution.

Salt

Salts are compounds that are formed whenever acids react. For example, when hydrochloric acid dissolves zinc, a salt called zinc chloride is formed.

Indicator

An indicator is a compound that changes colour depending upon whether it is in an acid or an alkali. Litmus (either as a solution or soaked into paper) is red in acid and blue in alkali.

Index